INTRODUCTION TO REMOTE SENSING

Edited by
Taras Kazantsev

Delve Publishing

Introduction to Remote Sensing

Edited by Taras Kazantsev

Edition 2018

Delve Publishing

Published by
Delve Publishing LLC
708 3rd Avenue, 6th Floor
New York
NY 10801
United States of America
www.delvepublishing.com

ISBN: 978-1-68095-668-9 (Hardback)

Printed in United States of America

Library of Congress Control Number: 2016920419

About the Editor

Taras Kazantsev

Taras Kazantsev has PhD in plant physiology since 2008. He is expert in remote sensing of vegetation and aerial mapping. He is authors of 15 papers and 2 patents. Besides research activity and articles writing, Taras Kazantsev is involved in a number of commercial projects on agromonitoring. Currently, Taras Kazantsev is research associate in Centre for Aero-Space research of Earth in Kyiv, Ukraine and head of Remote Sensing department in private company Drone.UA

Table of Contents

4 GLOBAL POSITIONING SYSTEM

5 HYPERSPECTRAL IMAGING

6 LIDAR APPLICATIONS

7 AERIAL PHOTOGRAPHY

8 REMOTE SENSING APPLICATION

List of Abbreviations

ABS	Anti-Lock Braking System
ACC	Adaptive Cruise Control
APN	Assessor's Parcel Number
AVHRR	Advanced Very High Resolution Radiometer
CAD	Computer Aided Design
CASA	Civil Aviation Safety Authority
CGIS	Canada Geographic Information System
CTIS	Computed Tomographic Imaging Spectrometry
EVM	Extraneous Vegetable Matter
FM	Foreign Material
FOC	Full Operational Capability
FSSD	Filter Stack Spectral Decomposition
GEO	Group on Earth Observations
GEOSS	Global Earth Observation System of Systems
GIS	Geographic Information System
GNSS	Global Navigation Satellite System
GPS	Global Positioning Systems
HSI	Hyperspectral Imaging
ICBM	Intercontinental Ballistic Missiles
IOC	Initial Operational Capability
IRIS	Image Replicating Imaging Spectrometry
LIDAR	Light Detection and Ranging
MIDAS	Mapping Display and Analysis System
MODIS	Moderate Resolution Imaging Spectroradiometer
MSI	Multispectral Sagnac interferometry
OBIA	Object Based Image Analysis
OCS	Operation Control Segment
OCX	Operational Control System
OGC	Open Geospacial Consortium
RDBMS	Relational Database Management System
SAL	Special Academic License

SAR	Synthetic Aperture Radar
SDK	Software Development Kits
SLBM	Submarine Launched Ballistic Missiles
TIN	Triangulated Irregular Networks
UAV	Unmanned Aerial Vehicles

Preface

Remote sensing is the art and science of recording, measuring, and analyzing information about a phenomenon from a distance. Humans with the aid of their eyes, noses, and ears are constantly seeing, smelling, and hearing things from a distance as they move through an environment. Thus, humans are naturally designed to be remote sensors. In order to study large areas of the Earth's surface geographers use devices known as remote sensors. These sensors are mounted on platforms such as helicopters, planes, and satellites that make it possible for the sensors to observe the Earth from above. Aircraft and satellites are the common platforms for remote sensing of the earth and its natural resources. Aerial photography in the visible portion of the electromagnetic wavelength was the original form of remote sensing but technological developments has enabled the acquisition of information at other wavelengths including near infrared, thermal infrared and microwave. Collection of information over a large numbers of wavelength bands is referred to as multispectral or hyper-spectral data. The development and deployment of manned and unmanned satellites has enhanced the collection of remotely sensed data and offers an inexpensive way to obtain information over large areas. The capacity of remote sensing to identify and monitor land surfaces and environmental conditions has expanded greatly over the last few years and remotely sensed data will be an essential tool in natural resource management. The era of satellite remote sensing was open in the 1960s when cameras and electronic sensors were mounted on spacecraft. Nowadays there is a big assortment of satellite systems actively recording information about the Earth. A wide variety of imagery is available from satellites. Both active and passive sensors, operating from the microwave to the ultraviolet regions of the electromagnetic spectrum collect a large amount of information about the earth's surface every day. Each of the systems vary in terms of their spatial, spectral, radiometic and temporal resolution.

Remote sensing makes use of visible, near infrared and short-wave infrared sensors to form images of the earth's surface by detecting the solar radiation reflected from targets on the ground. Different materials reflect and absorb differently at different wavelengths. Thus, the targets can be differentiated by their spectral reflectance signatures in the remotely sensed images. Remote sensing imagery has many applications in cartography, land use and cover, agriculture, soils mapping, forestry, city planning, grassland management, archaeology, military observations, meteorology, and geomorphology, among other uses. In order to use such imagery, one must have considerable knowledge about the Earth's surface and a strong background in remote sensing data acquisition and analysis techniques.

Since the Earth's surface consists of a mosaic of environmental conditions, geographers through their unique training in both the sciences and social sciences are well qualified to undertake various remote sensing applications.

Content Coverage

Chapters One and *Two* present an overview of remote sensing and GIS technology, correspondingly. GIS is a broad term that can refer to a number of different technologies, processes, and methods. It is attached to many operations and has many applications related to engineering, planning, management, transport/logistics, insurance, telecommunications, and business.

Chapters Three and *Four* present a coverage of earth observations satellite and GPS, consistently. Earth observation is the gathering of information about planet Earth's physical, chemical and biological systems, whereas The Global Positioning System (GPS) is a global navigation satellite system (GNSS) that provides geolocation and time information to a GPS receiver in all weather conditions, anywhere on or near the Earth where there is an unobstructed line of sight to four or more GPS satellites.

Chapters Five and *Six* are intended to focus on hyperspectral imaging and lidar, separately. The goal of hyperspectral imaging is to obtain the spectrum for each pixel in the image of a scene, with the purpose of finding objects, identifying materials, or detecting processes, while lidar is a surveying method that measures distance to a target by illuminating that target with a laser light.

Chapters Seven and *Eight* explain the fundamentals of aerial photography and remote sensing application, separately. A remote sensing application is a software application that processes remote sensing data. Remote sensing applications are similar to graphics software, but they enable generating geographic information from satellite and airborne sensor data.

CHAPTER 1

Remote Sensing: History and Scope

Chapter Contents

- **Introduction**
- **Overview**
- **Data Acquisition Techniques**
- **Applications of Remote Sensing**
- **Data Processing**
- **History**

Remote sensing refers to the activities of recording/observing/perceiving (sensing) objects or events at far away (remote) places. In remote sensing, the sensors are not in direct contact with the objects or events being observed. The information needs a physical carrier to travel from the objects/events to the sensors through an intervening medium. The electromagnetic radiation is normally used as an information carrier in remote sensing. The output of a remote sensing system is usually an image representing the scene being observed.

INTRODUCTION

Remote sensing is the acquisition of information about an object or phenomenon without making physical contact with the object and thus in contrast to on-site observation. Remote sensing is used in numerous fields, including geography and most Earth Science disciplines (for example, hydrology, ecology, oceanography, glaciology, geology); it also has military, intelligence, commercial, economic, planning, and humanitarian applications.

In current usage, the term "remote sensing" generally refers to the use of satellite- or aircraft-based sensor technologies to detect and classify objects on Earth, including on the surface and in the atmosphere and oceans, based on propagated signals (e.g. electromagnetic radiation). It may be split into "active" remote sensing (i.e., when a signal is emitted by a satellite or aircraft and its reflection by the object is detected by the sensor) and "passive" remote sensing (i.e., when the reflection of sunlight is detected by the sensor).

Synthetic aperture radar image of Death Valley colored using polarimetry.

OVERVIEW

Passive sensors gather radiation that is emitted or reflected by the object or surrounding areas. Reflected sunlight is the most common source of radiation measured by passive sensors. Examples of passive remote sensors include film photography, infrared, charge-coupled devices, and radiometers.

Active collection, on the other hand, emits energy in order to scan objects and areas whereupon a sensor then detects and measures the radiation that is reflected or backscattered from the target. RADAR and LiDAR are examples of active remote sensing where the time delay between emission and return is measured, establishing the location, speed and direction of an object.

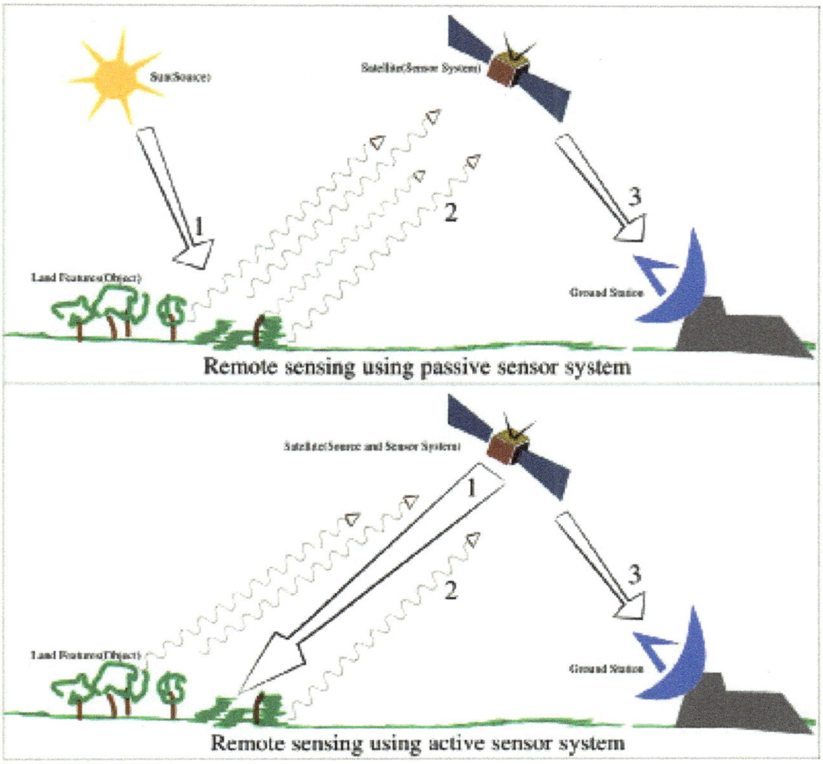

Illustration of Remote Sensing

Remote sensing makes it possible to collect data of dangerous or inaccessible areas. Remote sensing applications include monitoring deforestation in areas such as the Amazon Basin, glacial features in Arctic and Antarctic regions, and depth sounding of coastal and ocean depths. Military collection during the Cold War made use of stand-off collection of data about dangerous border areas. Remote sensing also replaces costly and slow data collection on the ground, ensuring in the process that areas or objects are not disturbed.

Orbital platforms collect and transmit data from different parts of the electromagnetic spectrum, which in conjunction with larger scale aerial or ground-based sensing and analysis, provides researchers with enough information to monitor trends such as El Niño and other natural long and short term phenomena. Other uses include different areas of the earth sciences such as natural resource management, agricultural fields such as land usage and conservation, and national security and overhead, ground-based and stand-off collection on border areas.

DATA ACQUISITION TECHNIQUES

The basis for multispectral collection and analysis is that of examined areas or objects that reflect or emit radiation that stand out from surrounding areas. For a summary of major remote sensing satellite systems see the overview table.

APPLICATIONS OF REMOTE SENSING

- Conventional radar is mostly associated with aerial traffic control, early warning, and certain large scale meteorological data. Doppler radar is used by local law enforcements' monitoring of speed limits and in enhanced meteorological collection such as wind speed and direction within weather systems in addition to precipitation location and intensity. Other types of active collection includes plasmas in the ionosphere. Interferometric synthetic aperture radar is used to produce precise digital elevation models of large scale terrain.
- Laser and radar altimeters on satellites have provided a wide range of data. By measuring the bulges of water caused by gravity, they map features on the seafloor to a resolution of a mile or so. By measuring the height and wavelength of ocean waves, the altimeters measure wind speeds and direction, and surface ocean currents and directions.
- Ultrasound (acoustic) and radar tide gauges measure sea level, tides and wave direction in coastal and offshore tide gauges.
- Light detection and ranging (LIDAR) is well known in examples of weapon ranging, laser illuminated homing of projectiles. LIDAR is used to detect and measure the concentration of various chemicals in the atmosphere, while airborne LIDAR can be used to measure heights of objects and features on the ground more accurately than with radar technology. Vegetation remote sensing is a principal application of LIDAR.
- Radiometers and photometers are the most common instrument in use, collecting reflected and emitted radiation in a wide range of frequencies. The most common are visible and infrared sensors, followed by microwave, gamma ray and rarely, ultraviolet. They may also be used to detect the emission spectra of various chemicals, providing data on chemical concentrations in the atmosphere.
- Stereographic pairs of aerial photographs have often been used to make topographic maps by imagery and terrain analysts in trafficability and highway departments for potential routes, in addition to modelling terrestrial habitat features.
- Simultaneous multi-spectral platforms such as Landsat have been in use since the 1970s. These thematic mappers take images in multiple wavelengths of electro-magnetic radiation (multi-spectral) and are usually found on Earth observation satellites, including (for example) the Landsat program or the IKONOS satellite. Maps of land cover and land use from thematic mapping can be used to prospect for minerals, detect or monitor land usage, detect invasive vegetation, deforestation, and examine the health of indigenous plants and crops, including entire farming regions or forests. Landsat images are used by regulatory agencies such as KYDOW to indicate water quality parameters including Secchi depth, chlorophyll a density and total phosphorus content. Weather satellites are used in meteorology and climatology.
- Hyperspectral imaging produces an image where each pixel has full spectral information with imaging narrow spectral bands over a contiguous spectral range. Hyperspectral imagers are used in various applications including mineralogy, biology, defense, and environmental measurements.

- Within the scope of the combat against desertification, remote sensing allows to follow up and monitor risk areas in the long term, to determine desertification factors, to support decision-makers in defining relevant measures of environmental management, and to assess their impacts.

Geodetic

- Geodetic remote sensing can be gravimetric or geometric. Overhead gravity data collection was first used in aerial submarine detection. This data revealed minute perturbations in the Earth's gravitational field that may be used to determine changes in the mass distribution of the Earth, which in turn may be used for geophysical studies, as in GRACE (satellite). Geometric remote sensing includes position and deformation imaging using InSAR, lidar, etc.

Acoustic and Near-Acoustic

- Sonar: *passive sonar*, listening for the sound made by another object (a vessel, a whale etc.); *active sonar*, emitting pulses of sounds and listening for echoes, used for detecting, ranging and measurements of underwater objects and terrain.
- Seismograms taken at different locations can locate and measure earthquakes (after they occur) by comparing the relative intensity and precise timings.
- Ultrasound: Ultrasound sensors, that emit high frequency pulses and listening for echoes, used for detecting water waves and water level, as in tide gauges or for towing tanks.

To coordinate a series of large-scale observations, most sensing systems depend on the following: platform location and the orientation of the sensor. High-end instruments now often use positional information from satellite navigation systems. The rotation and orientation is often provided within a degree or two with electronic compasses. Compasses can measure not just azimuth (i. e. degrees to magnetic north), but also altitude (degrees above the horizon), since the magnetic field curves into the Earth at different angles at different latitudes. More exact orientations require gyroscopic-aided orientation, periodically realigned by different methods including navigation from stars or known benchmarks.

DATA PROCESSING

Generally speaking, remote sensing works on the principle of the *inverse problem*. While the object or phenomenon of interest (the state) may not be directly measured, there exists some other variable that can be detected and measured (the observation) which may be related to the object of interest through a calculation. The common analogy given to describe this is trying to determine the type of animal from its footprints. For example, while it is impossible to directly measure temperatures in the upper atmosphere, it is possible to measure the spectral emissions from a known chemical species (such as carbon dioxide) in that region. The frequency of the emissions may then be related via thermodynamics to the temperature in that region.

The quality of remote sensing data consists of its spatial, spectral, radiometric and temporal resolutions.

Spatial Resolution

The size of a pixel that is recorded in a raster image – typically pixels may correspond to square areas ranging in side length from 1 to 1,000 meters (3.3 to 3,280.8 ft).

Spectral Resolution

The wavelength width of the different frequency bands recorded – usually, this is related to the number of frequency bands recorded by the platform. Current Landsat collection is that of seven bands, including several in the infra-red spectrum, ranging from a spectral resolution of 0.07 to 2.1 μm. The Hyperion sensor on Earth Observing-1 resolves 220 bands from 0.4 to 2.5 μm, with a spectral resolution of 0.10 to 0.11 μm per band.

Radiometric Resolution

The number of different intensities of radiation the sensor is able to distinguish. Typically, this ranges from 8 to 14 bits, corresponding to 256 levels of the gray scale and up to 16,384 intensities or "shades" of colour, in each band. It also depends on the instrument noise.

Temporal Resolution

The frequency of flyovers by the satellite or plane, and is only relevant in time-series studies or those requiring an averaged or mosaic image as in deforesting monitoring. This was first used by the intelligence community where repeated coverage revealed changes in infrastructure, the deployment of units or the modification/introduction of equipment. Cloud cover over a given area or object makes it necessary to repeat the collection of said location.

In order to create sensor-based maps, most remote sensing systems expect to extrapolate sensor data in relation to a reference point including distances between known points on the ground. This depends on the type of sensor used. For example, in conventional photographs, distances are accurate in the center of the image, with the distortion of measurements increasing the farther you get from the center. Another factor is that of the platen against which the film is pressed can cause severe errors when photographs are used to measure ground distances. The step in which this problem is resolved is called georeferencing, and involves computer-aided matching of points in the image (typically 30 or more points per image) which is extrapolated with the use of an established benchmark, "warping" the image to produce accurate spatial data. As of the early 1990s, most satellite images are sold fully georeferenced.

In addition, images may need to be radiometrically and atmospherically corrected.

Radiometric Correction

Allows to avoid radiometric errors and distortions. The illumination of objects on the Earth surface is uneven because of different properties of the relief. This factor is taken into account in the method of radiometric distortion correction. Radiometric correction gives a scale to the pixel values, e. g. the monochromatic scale of 0 to 255 will be converted to actual radiance values.

Topographic Correction (also called Terrain Correction)

In rugged mountains, as a result of terrain, the effective illumination of pixels varies considerably. In a remote sensing image, the pixel on the shady slope receives weak illumination and has a low radiance value, in contrast, the pixel on the sunny slope receives strong illumination and has a high radiance value. For the same object, the pixel radiance value on the shady slope will be different from that on the sunny slope. Additionally, different objects may have similar radiance values. These ambiguities seriously affected remote sensing image information extraction accuracy in mountainous areas. It became the main obstacle to further application of remote sensing images. The purpose of topographic correction is to eliminate this effect, recovering the true reflectivity or radiance of objects in horizontal conditions. It is the premise of quantitative remote sensing application.

Atmospheric Correction

Elimination of atmospheric haze by rescaling each frequency band so that its minimum value (usually realized in water bodies) corresponds to a pixel value of 0. The digitizing of data also makes it possible to manipulate the data by changing gray-scale values.

Interpretation is the critical process of making sense of the data. The first application was that of aerial photographic collection which used the following process; spatial measurement through the use of a light table in both conventional single or stereographic coverage, added skills such as the use of photogrammetry, the use of photomosaics, repeat coverage, Making use of objects' known dimensions in order to detect modifications. Image Analysis is the recently developed automated computer-aided application which is in increasing use.

Object-Based Image Analysis (OBIA) is a sub-discipline of GIS cience devoted to partitioning remote sensing (RS) imagery into meaningful image-objects, and assessing their characteristics through spatial, spectral and temporal scale.

Old data from remote sensing is often valuable because it may provide the only long-term data for a large extent of geography. At the same time, the data is often complex to interpret, and bulky to store. Modern systems tend to store the data digitally, often with lossless compression. The difficulty with this approach is that the data is fragile, the format may be archaic, and the data may be easy to falsify. One of the best systems for archiving data series is as computer-generated machine-readable ultrafiche, usually in typefonts such as OCR-B, or as digitized half-tone images. Ultrafiches survive well in standard libraries, with lifetimes of several centuries. They can be created, copied, filed and retrieved by automated systems. They are about as compact as archival magnetic media, and yet can be read by human beings with minimal, standardized equipment.

Data Processing Levels

To facilitate the discussion of data processing in practice, several processing "levels" were first defined in 1986 by NASA as part of its Earth Observing System and steadily adopted since then, both internally at NASA (e. g.,) and elsewhere (e. g.,); these definitions are:

Level	Description
0	Reconstructed, unprocessed instrument and payload data at full resolution, with any and all communications artifacts (e. g., synchronization frames, communications headers, duplicate data) removed.
1a	Reconstructed, unprocessed instrument data at full resolution, time-referenced, and annotated with ancillary information, including radiometric and geometric calibration coefficients and georeferencing parameters (e. g., platform ephemeris) computed and appended but not applied to the Level 0 data (or if applied, in a manner that level 0 is fully recoverable from level 1a data).
1b	Level 1a data that have been processed to sensor units (e. g., radar backscatter cross section, brightness temperature, etc.); not all instruments have Level 1b data; level 0 data is not recoverable from level 1b data.
2	Derived geophysical variables (e. g., ocean wave height, soil moisture, ice concentration) at the same resolution and location as Level 1 source data.
3	Variables mapped on uniform spacetime grid scales, usually with some completeness and consistency (e. g., missing points interpolated, complete regions mosaicked together from multiple orbits, etc.).
4	Model output or results from analyses of lower level data (i. e., variables that were not measured by the instruments but instead are derived from these measurements).

A Level 1 data record is the most fundamental (i. e., highest reversible level) data record that has significant scientific utility, and is the foundation upon which all subsequent data sets are produced. Level 2 is the first level that is directly usable for most scientific applications; its value is much greater than the lower levels. Level 2 data sets tend to be less voluminous than Level 1 data because they have been reduced temporally, spatially, or spectrally. Level 3 data sets are generally smaller than lower level data sets and thus can be dealt with without incurring a great deal of data handling overhead. These data tend to be generally more useful for many applications. The regular spatial and temporal organization of Level 3 datasets makes it feasible to readily combine data from different sources.

While these processing levels are particularly suitable for typical satellite data processing pipelines, other data level vocabularies have been defined and may be appropriate for more heterogeneous workflows.

HISTORY

The modern discipline of remote sensing arose with the development of flight. The balloonist G. Tournachon (alias Nadar) made photographs of Paris from his balloon in 1858. Messenger pigeons, kites, rockets and unmanned balloons were also used for early images. With the exception of balloons, these first, individual images were not particularly useful for map making or for scientific purposes.

The TR-1 reconnaissance/surveillance aircraft.

The *2001 Mars Odyssey* used spectrometers and imagers to hunt for evidence of past or present water and volcanic activity on Mars.

Systematic aerial photography was developed for military surveillance and reconnaissance purposes beginning in World War I and reaching a climax during the Cold War with the use of modified combat aircraft such as the P-51, P-38, RB-66 and the F-4C, or specifically designed collection platforms such as the U2/TR-1, SR-71, A-5 and the OV-1 series both in overhead and stand-off collection. A more recent development is that of increasingly smaller sensor pods such as those used by law enforcement and the military, in both manned and unmanned platforms. The advantage of this approach is that this requires minimal modification to a given airframe. Later imaging technologies would include Infra-red, conventional, Doppler and synthetic aperture radar.

The development of artificial satellites in the latter half of the 20th century allowed remote sensing to progress to a global scale as of the end of the Cold War. Instrumentation aboard various Earth observing and weather satellites such as Landsat, the Nimbus and more recent missions such as RADARSAT and UARS provided global measurements of various data for civil, research, and military purposes. Space probes to other planets have also provided the opportunity to conduct remote sensing studies in extraterrestrial environments, synthetic aperture radar aboard the Magellan spacecraft provided detailed topographic maps of Venus, while instruments aboard SOHO allowed studies to be performed on the Sun and the solar wind, just to name a few examples.

Recent developments include, beginning in the 1960s and 1970s with the development of image processing of satellite imagery. Several research groups in Silicon Valley including NASA Ames Research Center, GTE, and ESL Inc. developed Fourier transform techniques leading to the first notable enhancement of imagery data. In 1999 the first commercial satellite (IKONOS) collecting very high resolution imagery was launched.

Training and Education

At most universities remote sensing is associated with Geography departments. Remote Sensing has a growing relevance in the modern information society. It represents a key technology as part of the aerospace industry and bears increasing economic relevance – new sensors e.g. TerraSAR-X and RapidEye are developed constantly and the demand for skilled labour is increasing steadily. Furthermore, remote sensing exceedingly influences everyday life, ranging from weather forecasts to reports on climate change or natural disasters. As an example, 80% of the German students use the services of Google Earth; in 2006 alone the software was downloaded 100 million times. But studies have shown that only a fraction of them know more about the data they are working with. There exists a huge knowledge gap between the application and the understanding of satellite images. Remote sensing only plays a tangential role in schools, regardless of the political claims to strengthen the support for teaching on the subject. A lot of the computer software explicitly developed for school lessons has not yet been implemented due to its complexity. Thereby, the subject is either not at all integrated into the curriculum or does not pass the step of an interpretation of analogue images. In fact, the subject of remote sensing requires a consolidation of physics and mathematics as well as competences in the fields of media and methods apart from the mere visual interpretation of satellite images.

Many teachers have great interest in the subject "remote sensing", being motivated to integrate this topic into teaching, provided that the curriculum is considered. In many cases, this encouragement fails because of confusing information. In order to integrate remote sensing in a sustainable manner organizations like the EGU or digital earth encourages the development of learning modules and learning portals (e.g. FIS – Remote Sensing in School Lessons or Landmap – Spatial Discovery) promoting media and method qualifications as well as independent working.

Remote Sensing Software

Remote sensing data are processed and analyzed with computer software, known as a remote sensing application. A large number of proprietary and open source applications exist to process remote sensing data. Remote sensing software packages include:

- ERDAS IMAGINE from Hexagon Geospatial (Separated from Intergraph SG&I),
- PCI Geomatica
- TNTmips from MicroImages,
- IDRISI from Clark Labs,
- eCognition from Trimble,
- and RemoteView made by Overwatch Textron Systems.
- Dragon/ips is one of the oldest remote sensing packages still available, and is in some cases free.

Open source remote sensing software includes:

- Opticks (software),
- Orfeo toolbox
- Others mixing remote sensing and GIS capabilities are: GRASS GIS, ILWIS, QGIS, and TerraLook.

According to an NOAA Sponsored Research by Global Marketing Insights, Inc. the most used applications among Asian academic groups involved in remote sensing are as follows: ERDAS 36% (ERDAS IMAGINE 25% & ERMapper 11%); ESRI 30%; ITT Visual Information Solutions ENVI 17%; MapInfo 17%.

Among Western Academic respondents as follows: ESRI 39%, ERDAS IMAGINE 27%, MapInfo 9%, and AutoDesk 7%.

REFERENCES

1. Giacomo Capizzi, Grazia Lo Sciuto, Marcin Wozniak, Robertas Damasevicius: A Clustering Based System for Automated Oil Spill Detection by Satellite Remote Sensing, ICAISC (2) 2016: 613-623
2. Schowengerdt, Robert A. (2007). Remote sensing: models and methods for image processing (3rd ed.). Academic Press. p. 2. ISBN 978-0-12-369407-2.
3. Schott, John Robert (2007). Remote sensing: the image chain approach (2nd ed.). Oxford University Press. p. 1. ISBN 978-0-19-517817-3.

4. Guo, Huadong; Huang, Qingni; Li, Xinwu; Sun, Zhongchang; Zhang, Ying (2013). "Spatiotemporal analysis of urban environment based on the vegetation–impervious surface–soil model" (Full text article available). Journal of Applied Remote Sensing. 8: 084597. Bibcode:2014JARS....8.4597G. doi:10.1117/1.JRS.8.084597.
5. Liu, Jian Guo & Mason, Philippa J. (2009). Essential Image Processing for GIS and Remote Sensing. Wiley-Blackwell. p. 4. ISBN 978-0-470-51032-2.
6. "Saving the monkeys". SPIE Professional. Retrieved 1 Jan 2016.
7. Howard, A., et al., (Aug 19, 2015). "Remote sensing and habitat mapping for bearded capuchin monkeys (Sapajus libidinosus): landscapes for the use of stone tools". Journal of Applied Remote Sensing. 9 (1). doi:10.1117/1.JRS.9.096020.
8. http://hurricanes.nasa.gov/earth-sun/technology/remote_sensing.html
9. Mills, J.P.; et al. (1997). "Photogrammetry from Archived Digital Imagery for Seal Monitoring". The Photogrammetric Record. 15 (89): 715–724. doi:10.1111/0031-868X.00080.
10. Twiss, S.D.; et al. (2001). "Topographic spatial characterisation of grey seal Halichoerus grypus breeding habitat at a sub-seal size spatial grain". Ecography. 24 (3): 257–266. doi:10.1111/j.1600-0587.2001.tb00198.x.
11. Stewart, J.E.; et al. (2014). "Finescale ecological niche modeling provides evidence that lactating gray seals (Halichoerus grypus) prefer access to fresh water in order to drink". Marine Mammal Science. 30 (4): 1456–1472. doi:10.1111/mms.12126.
12. Begni G. Escadafal R. Fontannaz D. and Hong-Nga Nguyen A.-T. (2005). Remote sensing: a tool to monitor and assess desertification. Les dossiers thématiques du CSFD. Issue 2. 44 pp.
13. Geodetic Imaging
14. Grigoriev A.N. (2015). Method of radiometric distortion correction of multispectral data for the earth remote sensing.. Scientific and Technical Journal of Information Technologies, Mechanics and Optics. 15 (4): 595–602.
15. NASA (1986), Report of the EOS data panel, Earth Observing System, Data and Information System, Data Panel Report, Vol. IIa., NASA Technical Memorandum 87777, June 1986, 62 pp. Available at http://hdl.handle.net/2060/19860021622
16. C. L. Parkinson, A. Ward, M. D. King (Eds.) Earth Science Reference Handbook – A Guide to NASA's Earth Science Program and Earth Observing Satellite Missions, National Aeronautics and Space Administration Washington, D. C. Available at http://eospso.gsfc.nasa.gov/ftp_docs/2006ReferenceHandbook.pdf
17. GRAS-SAF (2009), Product User Manual, GRAS Satellite Application Facility, Version 1.2.1, 31 March 2009. Available at http://www.grassaf.org/general-documents/products/grassaf_pum_v121.pdf
18. Ditter, R., Haspel, M., Jahn, M., Kollar, I., Siegmund, A., Viehrig, K., Volz, D., Siegmund, A. (2012) Geospatial technologies in school – theoretical concept and practical implementation in K-12 schools. In: International Journal of Data

Mining, Modelling and Management (IJDMMM): FutureGIS: Riding the Wave of a Growing Geospatial Technology Literate Society; Vol. X

19. Stork, E.J., Sakamoto, S.O., and Cowan, R.M. (1999) The integration of science explorations through the use of earth images in middle school curriculum, Proc. IEEE Trans. Geosci. Remote Sensing 37, 1801–1817
20. Bednarz, S.W. and Whisenant, S.E. (2000) Mission geography: linking national geography standards, innovative technologies and NASA, Proc. IGARSS, Honolulu, USA, 2780–2782

CHAPTER 2

GIS Technology And Applications

Chapter Contents

- **Introduction**
- **History of Development**
- **GIS Techniques and Technology**
- **Spatial Analysis with Geographical Information System (GIS)**
- **Applications**
- **GIS File Formats**
- **Map Projection**
- **Geocoding**
- **Web Mapping**
- **Geographic Information Systems in Geospatial Intelligence**

A geographic information system (GIS) is a computer system for capturing, storing, checking, and displaying data related to positions on Earth's surface. GIS can show many different kinds of data on one map. This enables people to more easily see, analyze, and understand patterns and relationships.

With GIS technology, people can compare the locations of different things in order to discover how they relate to each other. For example, using GIS, the same map could include sites that produce pollution, such as gas stations, and sites that are sensitive to pollution, such as wetlands. Such a map would help people determine which wetlands are most at risk.

INTRODUCTION

A geographic information system (or GIS) is a system designed to capture, store, manipulate, analyze, manage, and present spatial or geographical data. The acronym GIS is sometimes used for geographic information science (GIS cience) to refer to the academic discipline that studies geographic information systems and is a large domain within the broader academic discipline of geoinformatics. What goes beyond a GIS is a spatial data infrastructure, a concept that has no such restrictive boundaries.

In general, the term describes any information system that integrates, stores, edits, analyzes, shares, and displays geographic information. GIS applications are tools that allow users to create interactive queries (user-created searches), analyze spatial information, edit data in maps, and present the results of all these operations. Geographic information science is the science underlying geographic concepts, applications, and systems.

GIS is a broad term that can refer to a number of different technologies, processes, and methods. It is attached to many operations and has many applications related to engineering, planning, management, transport/logistics, insurance, telecommunications, and business. For that reason, GIS and location intelligence applications can be the foundation for many location-enabled services that rely on analysis and visualization.

GIS can relate unrelated information by using location as the key index variable. Locations or extents in the Earth space–time may be recorded as dates/times of occurrence, and x, y, and z coordinates representing, longitude, latitude, and elevation, respectively. All Earth-based spatial–temporal location and extent references should be relatable to one another and ultimately to a "real" physical location or extent. This key characteristic of GIS has begun to open new avenues of scientific inquiry.

HISTORY OF DEVELOPMENT

The first known use of the term "geographic information system" was by Roger Tomlinson in the year 1968 in his paper "A Geographic Information System for Regional Planning". Tomlinson is also acknowledged as the "father of GIS".

E. W. Gilbert's version (1958) of John Snow's 1855 map of the Soho cholera outbreak showing the clusters of cholera cases in the London epidemic of 1854

Previously, one of the first applications of spatial analysis in epidemiology is the 1832 "Rapport sur la marche et les effets du choléra dans Paris et le département de la Seine". The French geographer Charles Picquet represented the 48 districts of the city of Paris by halftone color gradient according to the number of deaths by cholera per 1,000 inhabitants. In 1854 John Snow determined the source of a cholera outbreak in London by marking points on a map depicting where the cholera victims lived, and connecting the cluster that he found with a nearby water source. This was one of the earliest successful uses of a geographic methodology in epidemiology. While the basic elements of topography and theme existed previously in cartography, the John Snow map was unique, using cartographic methods not only to depict but also to analyze clusters of geographically dependent phenomena.

The early 20th century saw the development of photozincography, which allowed maps to be split into layers, for example one layer for vegetation and another for water. This was particularly used for printing contours – drawing these was a labor-intensive task but having them on a separate layer meant they could be worked on without the other layers to confuse the draughtsman. This work was originally drawn on glass plates but later plastic film was introduced, with the advantages of being lighter, using less storage space and being less brittle, among others. When all the layers were finished, they were combined into one image using a large process camera. Once color printing came in, the layers idea was also used for creating separate printing plates for each color. While the use of layers much later became one of the main typical features of a contemporary GIS, the photographic process just described is not considered to be a GIS in itself – as the maps were just images with no database to link them to.

Computer hardware development spurred by nuclear weapon research led to general-purpose computer "mapping" applications by the early 1960s.

The year 1960 saw the development of the world's first true operational GIS in Ottawa, Ontario, Canada by the federal Department of Forestry and Rural Development. Developed by Dr. Roger Tomlinson, it was called the Canada Geographic Information System (CGIS) and was used to store, analyze, and manipulate data collected for the Canada Land Inventory – an effort to determine the land capability for rural Canada by mapping information about soils, agriculture, recreation, wildlife, waterfowl, forestry and land use at a scale of 1:50,000. A rating classification factor was also added to permit analysis.

CGIS was an improvement over "computer mapping" applications as it provided capabilities for overlay, measurement, and digitizing/scanning. It supported a national coordinate system that spanned the continent, coded lines as arcs having a true embedded topology and it stored the attribute and locational information in separate files. As a result of this, Tomlinson has become known as the "father of GIS", particularly for his use of overlays in promoting the spatial analysis of convergent geographic data.

CGIS lasted into the 1990s and built a large digital land resource database in Canada. It was developed as a mainframe-based system in support of federal and provincial resource planning and management. Its strength was continent-wide analysis of complex datasets. The CGIS was never available commercially.

In 1964 Howard T. Fisher formed the Laboratory for Computer Graphics and Spatial Analysis at the Harvard Graduate School of Design (LCGSA 1965–1991), where a number of important theoretical concepts in spatial data handling were developed, and which by the 1970s had distributed seminal software code and systems, such as SYMAP, GRID, and ODYSSEY – that served as sources for subsequent commercial development—to universities, research centers and corporations worldwide.

By the late 1970s two public domain GIS systems (MOSS and GRASS GIS) were in development, and by the early 1980s, M&S Computing (later Intergraph) along with Bentley Systems Incorporated for the CAD platform, Environmental Systems Research Institute (ESRI), CARIS (Computer Aided Resource Information System), MapInfo Corporation and ERDAS (Earth Resource Data Analysis System) emerged as commercial vendors of GIS software, successfully incorporating many of the CGIS features, combining the first generation approach to separation of spatial and attribute information with a second generation approach to organizing attribute data into database structures.

In 1986, Mapping Display and Analysis System (MIDAS), the first desktop GIS product was released for the DOS operating system. This was renamed in 1990 to MapInfo for Windows when it was ported to the Microsoft Windows platform. This began the process of moving GIS from the research department into the business environment.

By the end of the 20th century, the rapid growth in various systems had been consolidated and standardized on relatively few platforms and users were beginning to explore viewing GIS data over the Internet, requiring data format and transfer standards. More recently, a growing number of free, open-source GIS packages run on a range of operating systems and can be customized to perform specific tasks. Increasingly geospatial data and mapping applications are being made available via the world wide web.

GIS TECHNIQUES AND TECHNOLOGY

Modern GIS technologies use digital information, for which various digitized data creation methods are used. The most common method of data creation is digitization, where a hard copy map or survey plan is transferred into a digital medium through the use of a CAD program, and geo-referencing capabilities. With the wide availability of ortho-rectified imagery (from satellites, aircraft, Helikites and UAVs), heads-up digitizing is becoming the main avenue through which geographic data is extracted. Heads-up digitizing involves the tracing of geographic data directly on top of the aerial imagery instead of by the traditional method of tracing the geographic form on a separate digitizing tablet (heads-down digitizing).

Relating Information from Different Sources

GIS uses spatio-temporal (space-time) location as the key index variable for all other information. Just as a relational database containing text or numbers can relate many different tables using common key index variables, GIS can relate otherwise unrelated information by using location as the key index variable. The key is the location and/or extent in space-time.

Any variable that can be located spatially, and increasingly also temporally, can be referenced using a GIS. Locations or extents in Earth space–time may be recorded as dates/times of occurrence, and x, y, and z coordinates representing, longitude, latitude, and elevation, respectively. These GIS coordinates may represent other quantified systems of temporo-spatial reference (for example, film frame number, stream gage station, highway mile-marker, surveyor benchmark, building address, street intersection, entrance gate, water depth sounding, POS or CAD drawing origin/units). Units applied to recorded temporal-spatial data can vary widely (even when using exactly the same data, see map projections), but all Earth-based spatial–temporal location and extent references should, ideally, be relatable to one another and ultimately to a "real" physical location or extent in space–time.

Related by accurate spatial information, an incredible variety of real-world and projected past or future data can be analyzed, interpreted and represented. This key characteristic of GIS has begun to open new avenues of scientific inquiry into behaviors and patterns of real-world information that previously had not been systematically correlated.

GIS Uncertainties

GIS accuracy depends upon source data, and how it is encoded to be data referenced. Land surveyors have been able to provide a high level of positional accuracy utilizing the GPS-derived positions. High-resolution digital terrain and aerial imagery, powerful computers and Web technology are changing the quality, utility, and expectations of GIS to serve society on a grand scale, but nevertheless there are other source data that affect overall GIS accuracy like paper maps, though these may be of limited use in achieving the desired accuracy.

In developing a digital topographic database for a GIS, topographical maps are the main source, and aerial photography and satellite imagery are extra sources for collecting data and identifying attributes which can be mapped in layers over a location facsimile of scale. The scale of a map and geographical rendering area representation type are very important aspects since the information content depends mainly on the scale set and resulting locatability of the map's representations. In order to digitize a map, the map has to be checked within theoretical dimensions, then scanned into a raster format, and resulting raster data has to be given a theoretical dimension by a rubber sheeting/warping technology process.

A quantitative analysis of maps brings accuracy issues into focus. The electronic and other equipment used to make measurements for GIS is far more precise than the machines of conventional map analysis. All geographical data are inherently inaccurate, and these inaccuracies will propagate through GIS operations in ways that are difficult to predict.

Data Representation

GIS data represents real objects (such as roads, land use, elevation, trees, waterways, etc.) with digital data determining the mix. Real objects can be divided into two abstractions: discrete objects (e.g., a house) and continuous fields (such as rainfall amount, or elevations).

Traditionally, there are two broad methods used to store data in a GIS for both kinds of abstractions mapping references: raster images and vector. Points, lines, and polygons are the stuff of mapped location attribute references. A new hybrid method of storing data is that of identifying point clouds, which combine three-dimensional points with RGB information at each point, returning a "3D color image". GIS thematic maps then are becoming more and more realistically visually descriptive of what they set out to show or determine.

Data Capture

Existing data printed on paper or PET film maps can be digitized or scanned to produce digital data. A digitizer produces vector data as an operator traces points, lines, and polygon boundaries from a map. Scanning a map results in raster data that could be further processed to produce vector data.

Example of hardware for mapping (GPS and laser rangefinder) and data collection (rugged computer). The current trend for geographical information system (GIS) is that accurate mapping and data analysis are completed while in the field. Depicted hardware (field-map technology) is used mainly for forest inventories, monitoring and mapping.

Data capture—entering information into the system—consumes much of the time of GIS practitioners. There are a variety of methods used to enter data into a GIS where it is stored in a digital format.

Survey data can be directly entered into a GIS from digital data collection systems on survey instruments using a technique called coordinate geometry (COGO). Positions from a global navigation satellite system (GNSS) like Global Positioning System can also be collected and then imported into a GIS. A current trend in data collection gives users the ability to utilize field computers with the ability to edit live data using wireless connections or disconnected editing sessions. This has been enhanced by the availability of low-cost mapping-grade GPS units with decimeter accuracy in real time. This eliminates the need to post process, import, and update the data in the office after fieldwork has been collected. This includes the ability to incorporate positions collected using a laser rangefinder. New technologies also allow users to create maps as well as analysis directly in the field, making projects more efficient and mapping more accurate.

Remotely sensed data also plays an important role in data collection and consist of sensors attached to a platform. Sensors include cameras, digital scanners and lidar, while platforms usually consist of aircraft and satellites. In England in the mid 1990s, hybrid kite/balloons called Helikites first pioneered the use of compact airborne digital cameras as airborne Geo-Information Systems. Aircraft measurement software, accurate to 0.4 mm was used to link the photographs and measure the ground. Helikites are inexpensive and gather more accurate data than aircraft. Helikites can be used over roads, railways and towns where UAVs are banned.

Recently with the development of miniature UAVs, aerial data collection is becoming possible with them. For example, the Aeryon Scout was used to map a 50-acre area with a Ground sample distance of 1 inch (2.54 cm) in only 12 minutes.

The majority of digital data currently comes from photo interpretation of aerial photographs. Soft-copy workstations are used to digitize features directly from stereo pairs of digital photographs. These systems allow data to be captured in two and three dimensions, with elevations measured directly from a stereo pair using principles of photogrammetry. Analog aerial photos must be scanned before being entered into a soft-copy system, for high-quality digital cameras this step is skipped.

Satellite remote sensing provides another important source of spatial data. Here satellites use different sensor packages to passively measure the reflectance from parts of the electromagnetic spectrum or radio waves that were sent out from an active sensor such as radar. Remote sensing collects raster data that can be further processed using different bands to identify objects and classes of interest, such as land cover.

When data is captured, the user should consider if the data should be captured with either a relative accuracy or absolute accuracy, since this could not only influence how information will be interpreted but also the cost of data capture.

After entering data into a GIS, the data usually requires editing, to remove errors, or further processing. For vector data it must be made "topologically correct" before it can be used for some advanced analysis. For example, in a road network, lines must connect with nodes at an intersection. Errors such as undershoots and overshoots must also be removed. For scanned maps, blemishes on the source map may need to be removed from the resulting raster. For example, a fleck of dirt might connect two lines that should not be connected.

Raster-to-Vector Translation

Data restructuring can be performed by a GIS to convert data into different formats. For example, a GIS may be used to convert a satellite image map to a vector structure by generating lines around all cells with the same classification, while determining the cell spatial relationships, such as adjacency or inclusion.

More advanced data processing can occur with image processing, a technique developed in the late 1960s by NASA and the private sector to provide contrast enhancement, false color rendering and a variety of other techniques including use of two dimensional Fourier transforms. Since digital data is collected and stored in various ways, the two data sources may not be entirely compatible. So a GIS must be able to convert geographic data from one structure to another. In so doing, the implicit assumptions behind different ontologies and classifications require analysis. Object ontologies have gained increasing prominence as a consequence of object-oriented programming and sustained work by Barry Smith and co-workers.

Projections, Coordinate Systems, and Registration

The earth can be represented by various models, each of which may provide a different set of coordinates (e.g., latitude, longitude, elevation) for any given point on the Earth's surface. The simplest model is to assume the earth is a perfect sphere. As more measurements of the earth have accumulated, the models of the earth have become more sophisticated and more accurate. In fact, there are models called datums that apply to different areas of the earth to provide increased accuracy, like NAD83 for U.S. measurements, and the World Geodetic System for worldwide measurements.

SPATIAL ANALYSIS WITH GEOGRAPHICAL INFORMATION SYSTEM (GIS)

GIS spatial analysis is a rapidly changing field, and GIS packages are increasingly including analytical tools as standard built-in facilities, as optional toolsets, as add-ins or 'analysts'. In many instances these are provided by the original software suppliers (commercial vendors or collaborative non commercial development teams), while in other cases facilities have been developed and are provided by third parties. Furthermore, many products offer software development kits (SDKs), programming languages and language support, scripting facilities and/or special interfaces for developing one's own analytical tools or variants. The website "Geospatial Analysis" and associated book/ebook attempt to provide a reasonably comprehensive guide to the subject. The increased availability has created a new dimension to business intelligence termed "spatial intelligence" which, when openly delivered via intranet, democratizes access to geographic and social network data. Geospatial intelligence, based on GIS spatial analysis, has also become a key element for security. GIS as a whole can be described as conversion to a vectorial representation or to any other digitization process.

Slope and Aspect

Slope can be defined as the steepness or gradient of a unit of terrain, usually measured as an angle in degrees or as a percentage. Aspect can be defined as the direction in which a unit of terrain faces. Aspect is usually expressed in degrees from north. Slope, aspect, and surface curvature in terrain analysis are all derived from neighborhood operations using elevation values of a cell's adjacent neighbors. Slope is a function of resolution, and the spatial resolution used to calculate slope and aspect should always be specified. Authors such as Skidmore, Jones and Zhou and Liu have compared techniques for calculating slope and aspect.

The following method can be used to derive slope and aspect:

The elevation at a point or unit of terrain will have perpendicular tangents (slope) passing through the point, in an east-west and north-south direction. These two tangents give two components, $\partial z/\partial x$ and $\partial z/\partial y$, which then be used to determine the overall direction of slope, and the aspect of the slope. The gradient is defined as a vector quantity with components equal to the partial derivatives of the surface in the x and y directions.

The calculation of the overall 3x3 grid slope S and aspect A for methods that determine east-west and north-south component use the following formulas respectively:

$$\tan S = \sqrt{\left(\frac{\partial z}{\partial x}\right)^2 + \left(\frac{\partial z}{\partial y}\right)^2}$$

$$\tan A = \left(\frac{\left(\frac{-\partial z}{\partial y}\right)}{\left(\frac{\partial z}{\partial x}\right)}\right)$$

Zhou and Liu describe another formula for calculating aspect, as follows:

$$A = 270^\circ + \arctan\left(\frac{\left(\frac{\partial z}{\partial x}\right)}{\left(\frac{\partial z}{\partial y}\right)}\right) - 90^\circ \left(\frac{\left(\frac{\partial z}{\partial y}\right)}{\left|\frac{\partial z}{\partial y}\right|}\right)$$

Data Analysis

It is difficult to relate wetlands maps to rainfall amounts recorded at different points such as airports, television stations, and schools. A GIS, however, can be used to depict two- and three-dimensional characteristics of the Earth's surface, subsurface, and atmosphere from information points. For example, a GIS can quickly generate a map with isopleth or contour lines that indicate differing amounts of rainfall. Such a map can be thought of as a rainfall contour map. Many sophisticated methods can estimate the characteristics of surfaces from a limited number of point measurements. A two-dimensional contour map created from the surface modeling of rainfall point measurements may be overlaid and analyzed with any other map in a GIS covering the same area. This GIS derived map can then provide additional information - such as the viability of water power potential as a renewable energy source. Similarly, GIS can be used to compare other renewable energy resources to find the best geographic potential for a region.

Additionally, from a series of three-dimensional points, or digital elevation model, isopleth lines representing elevation contours can be generated, along with slope analysis, shaded relief, and other elevation products. Watersheds can be easily defined for any given reach, by computing all of the areas contiguous and uphill from any given point of interest. Similarly, an expected thalweg of where surface water would want to travel in intermittent and permanent streams can be computed from elevation data in the GIS.

Topological Modeling

A GIS can recognize and analyze the spatial relationships that exist within digitally stored spatial data. These topological relationships allow complex spatial modelling and analysis to be performed. Topological relationships between geometric entities traditionally include adjacency (what adjoins what), containment (what encloses what), and proximity (how close something is to something else).

Geometric Networks

Geometric networks are linear networks of objects that can be used to represent interconnected features, and to perform special spatial analysis on them. A geometric network is composed of edges, which are connected at junction points, similar to graphs in mathematics and computer science. Just like graphs, networks can have weight and flow assigned to its edges, which can be used to represent various interconnected features more accurately. Geometric networks are often used to model road networks and public utility networks, such as electric, gas, and water networks. Network modeling is also commonly employed in transportation planning, hydrology modeling, and infrastructure modeling.

Hydrological Modeling

GIS hydrological models can provide a spatial element that other hydrological models lack, with the analysis of variables such as slope, aspect and watershed or catchment area. Terrain analysis is fundamental to hydrology, since water always flows down a slope. As basic terrain analysis of a digital elevation model (DEM) involves calculation of slope and aspect, DEMs are very useful for hydrological analysis. Slope and aspect can then be used to determine direction of surface runoff, and hence flow accumulation for the formation of streams, rivers and lakes. Areas of divergent flow can also give a clear indication of the boundaries of a catchment. Once a flow direction and accumulation matrix has been created, queries can be performed that show contributing or dispersal areas at a certain point. More detail can be added to the model, such as terrain roughness, vegetation types and soil types, which can influence infiltration and evapotranspiration rates, and hence influencing surface flow. One of the main uses of hydrological modeling is in environmental contamination research.

Cartographic Modeling

The term "cartographic modeling" was probably coined by Dana Tomlin in his PhD dissertation and later in his book which has the term in the title.

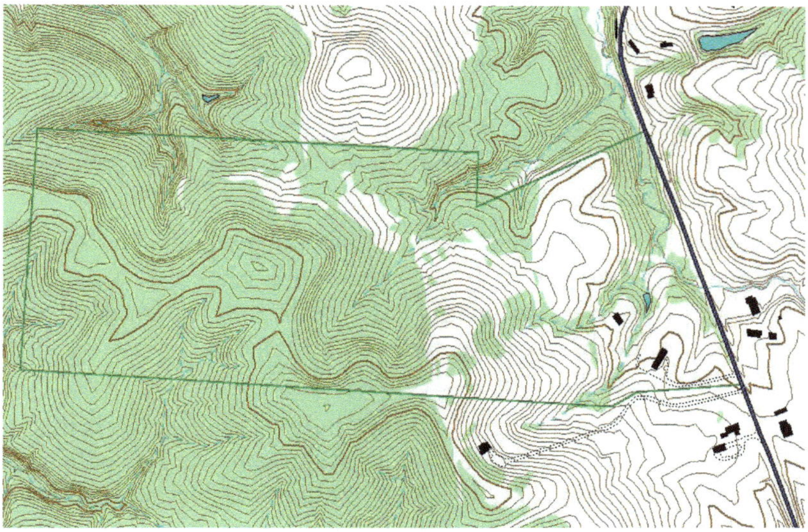

An example of use of layers in a GIS application. In this example, the forest cover layer (light green) is at the bottom, with the topographic layer over it. Next up is the stream layer, then the boundary layer, then the road layer. The order is very important in order to properly display the final result. Note that the pond layer was located just below the stream layer, so that a stream line can be seen overlying one of the ponds.

Cartographic modeling refers to a process where several thematic layers of the same area are produced, processed, and analyzed. Tomlin used raster layers, but the overlay method (see below) can be used more generally. Operations on map layers can be combined into algorithms, and eventually into simulation or optimization models.

Map Overlay

The combination of several spatial datasets (points, lines, or polygons) creates a new output vector dataset, visually similar to stacking several maps of the same region. These overlays are similar to mathematical Venn diagram overlays. A union overlay combines the geographic features and attribute tables of both inputs into a single new output. An intersect overlay defines the area where both inputs overlap and retains a set of attribute fields for each. A symmetric difference overlay defines an output area that includes the total area of both inputs except for the overlapping area.

Data extraction is a GIS process similar to vector overlay, though it can be used in either vector or raster data analysis. Rather than combining the properties and features of both datasets, data extraction involves using a "clip" or "mask" to extract the features of one data set that fall within the spatial extent of another dataset.

In raster data analysis, the overlay of datasets is accomplished through a process known as "local operation on multiple rasters" or "map algebra," through a function that combines the values of each raster's matrix. This function may weigh some inputs more than others through use of an "index model" that reflects the influence of various factors upon a geographic phenomenon.

Geostatistics

Geostatistics is a branch of statistics that deals with field data, spatial data with a continuous index. It provides methods to model spatial correlation, and predict values at arbitrary locations (interpolation).

When phenomena are measured, the observation methods dictate the accuracy of any subsequent analysis. Due to the nature of the data (e.g. traffic patterns in an urban environment; weather patterns over the Pacific Ocean), a constant or dynamic degree of precision is always lost in the measurement. This loss of precision is determined from the scale and distribution of the data collection.

To determine the statistical relevance of the analysis, an average is determined so that points (gradients) outside of any immediate measurement can be included to determine their predicted behavior. This is due to the limitations of the applied statistic and data collection methods, and interpolation is required to predict the behavior of particles, points, and locations that are not directly measurable.

Hillshade model derived from a Digital Elevation Model of the Valestra area in the northern Apennines (Italy)

Interpolation is the process by which a surface is created, usually a raster dataset, through the input of data collected at a number of sample points. There are several forms of interpolation, each which treats the data differently, depending on the properties of the data set. In comparing interpolation methods, the first consideration should be whether or not the source data will change (exact or approximate). Next is whether the method is subjective, a human interpretation, or objective. Then there is the nature of transitions between points: are they abrupt or gradual. Finally, there is whether a method is global (it uses the entire data set to form the model), or local where an algorithm is repeated for a small section of terrain.

Interpolation is a justified measurement because of a spatial autocorrelation principle that recognizes that data collected at any position will have a great similarity to, or influence of those locations within its immediate vicinity.

Digital elevation models, triangulated irregular networks, edge-finding algorithms, Thiessen polygons, Fourier analysis, (weighted) moving averages, inverse distance weighting, kriging, spline, and trend surface analysis are all mathematical methods to produce interpolative data.

Address Geocoding

Geocoding is interpolating spatial locations (X,Y coordinates) from street addresses or any other spatially referenced data such as ZIP Codes, parcel lots and address locations. A reference theme is required to geocode individual addresses, such as a road centerline file with address ranges. The individual address locations have historically been interpolated, or estimated, by examining address ranges along a road segment. These are usually provided in the form of a table or database. The software will then place a dot approximately where that address belongs along the segment of centerline. For example, an address point of 500 will be at the midpoint of a line segment that starts with address 1 and ends with address 1,000. Geocoding can also be applied against actual parcel data, typically from municipal tax maps. In this case, the result of the geocoding will be an actually positioned space as opposed to an interpolated point. This approach is being increasingly used to provide more precise location information.

Reverse Geocoding

Reverse geocoding is the process of returning an estimated street address number as it relates to a given coordinate. For example, a user can click on a road centerline theme (thus providing a coordinate) and have information returned that reflects the estimated house number. This house number is interpolated from a range assigned to that road segment. If the user clicks at the midpoint of a segment that starts with address 1 and ends with 100, the returned value will be somewhere near 50. Note that reverse geocoding does not return actual addresses, only estimates of what should be there based on the predetermined range.

Multi-Criteria Decision Analysis

Coupled with GIS, multi-criteria decision analysis methods support decision-makers in analyzing a set of alternative spatial solutions, such as the most likely ecological habitat for restoration, against multiple criteria, such as vegetation cover or roads. MCDA uses decision rules to aggregate the criteria, which allows the alternative solutions to be ranked or prioritized. GIS MCDA may reduce costs and time involved in identifying potential restoration sites.

Data Output and Cartography

Cartography is the design and production of maps, or visual representations of spatial data. The vast majority of modern cartography is done with the help of computers, usually using GIS but production of quality cartography is also achieved by importing layers into a design program to refine it. Most GIS software gives the user substantial control over the appearance of the data.

Cartographic work serves two major functions:

First, it produces graphics on the screen or on paper that convey the results of analysis to the people who make decisions about resources. Wall maps and other graphics can be generated, allowing the viewer to visualize and thereby understand the results of analyses or simulations of potential events. Web Map Servers facilitate distribution of generated maps through web browsers using various implementations of web-based application programming interfaces (AJAX, Java, Flash, etc.).

Second, other database information can be generated for further analysis or use. An example would be a list of all addresses within one mile (1.6 km) of a toxic spill.

Graphic Display Techniques

Traditional maps are abstractions of the real world, a sampling of important elements portrayed on a sheet of paper with symbols to represent physical objects. People who use maps must interpret these symbols. Topographic maps show the shape of land surface with contour lines or with shaded relief.

Today, graphic display techniques such as shading based on altitude in a GIS can make relationships among map elements visible, heightening one's ability to extract and analyze information. For example, two types of data were combined in a GIS to produce a perspective view of a portion of San Mateo County, California.

- The digital elevation model, consisting of surface elevations recorded on a 30-meter horizontal grid, shows high elevations as white and low elevation as black.
- The accompanying Landsat Thematic Mapper image shows a false-color infrared image looking down at the same area in 30-meter pixels, or picture elements, for the same coordinate points, pixel by pixel, as the elevation information.

A GIS was used to register and combine the two images to render the three-dimensional perspective view looking down the San Andreas Fault, using the Thematic Mapper image pixels, but shaded using the elevation of the landforms. The GIS display depends on the viewing point of the observer and time of day of the display, to properly render the shadows created by the sun's rays at that latitude, longitude, and time of day.

An archeochrome is a new way of displaying spatial data. It is a thematic on a 3D map that is applied to a specific building or a part of a building. It is suited to the visual display of heat-loss data.

Spatial ETL

Spatial ETL tools provide the data processing functionality of traditional Extract, Transform, Load (ETL) software, but with a primary focus on the ability to manage spatial data. They provide GIS users with the ability to translate data between different standards and proprietary formats, whilst geometrically transforming the data en route. These tools can come in the form of add-ins to existing wider-purpose software such as Microsoft Excel.

GIS Data Mining

GIS or spatial data mining is the application of data mining methods to spatial data. Data mining, which is the partially automated search for hidden patterns in large databases, offers great potential benefits for applied GIS-based decision making. Typical applications include environmental monitoring. A characteristic of such applications is that spatial correlation between data measurements require the use of specialized algorithms for more efficient data analysis.

APPLICATIONS

The implementation of a GIS is often driven by jurisdictional (such as a city), purpose, or application requirements. Generally, a GIS implementation may be custom-designed for an organization. Hence, a GIS deployment developed for an application, jurisdiction, enterprise, or purpose may not be necessarily interoperable or compatible with a GIS that has been developed for some other application, jurisdiction, enterprise, or purpose.

GIS provides, for every kind of location-based organization, a platform to update geographical data without wasting time to visit the field and update a database manually. GIS when integrated with other powerful enterprise solutions like SAP and the Wolfram Language helps creating powerful decision support system at enterprise level.

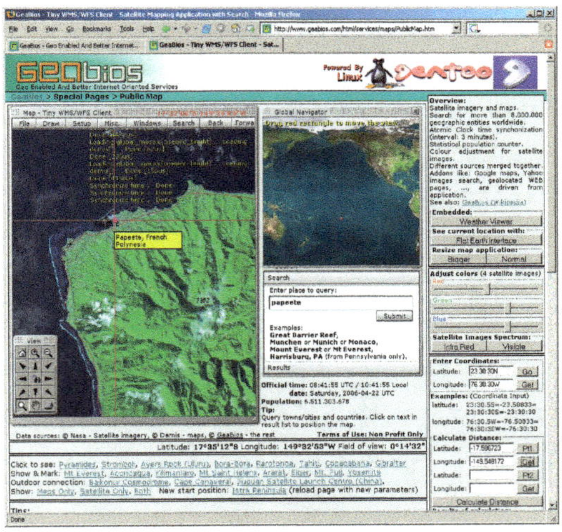

GeaBios – tiny WMS/WFS client (Flash/DHTML)

Many disciplines can benefit from GIS technology. An active GIS market has resulted in lower costs and continual improvements in the hardware and software components of GIS, and usage in the fields of science, government, business, and industry, with applications including real estate, public health, crime mapping, national defense, sustainable development, natural resources, climatology, landscape architecture, archaeology, regional and community planning, transportation and logistics. GIS is also diverging into location-based services, which allows GPS-enabled mobile devices to display their location in relation to fixed objects (nearest restaurant, gas station, fire hydrant) or mobile objects (friends, children, police car), or to relay their position back to a central server for display or other processing.

Open Geospatial Consortium standards

The Open Geospatial Consortium (OGC) is an international industry consortium of 384 companies, government agencies, universities, and individuals participating in a consensus process to develop publicly available geoprocessing specifications. Open interfaces and protocols defined by OpenGIS Specifications support interoperable solutions that "geo-enable" the Web, wireless and location-based services, and mainstream IT, and empower technology developers to make complex spatial information and services accessible and useful with all kinds of applications. Open Geospatial Consortium protocols include Web Map Service, and Web Feature Service.

GIS products are broken down by the OGC into two categories, based on how completely and accurately the software follows the OGC specifications.

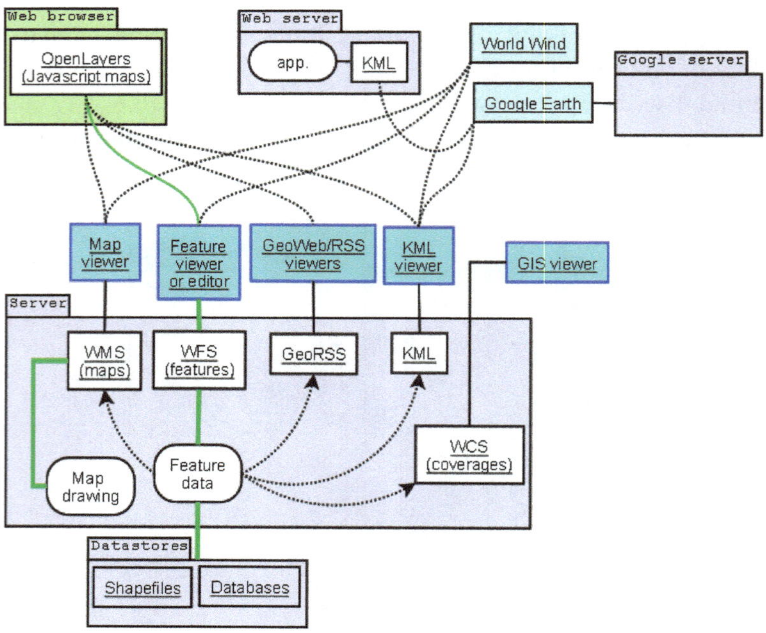

OGC standards help GIS tools communicate.

Compliant Products are software products that comply to OGC's OpenGIS Specifications. When a product has been tested and certified as compliant through the OGC Testing Program, the product is automatically registered as "compliant" on this site.

Implementing Products are software products that implement OpenGIS Specifications but have not yet passed a compliance test. Compliance tests are not available for all specifications. Developers can register their products as implementing draft or approved specifications, though OGC reserves the right to review and verify each entry.

Web Mapping

In recent years there has been an proliferation of free-to-use and easily accessible mapping software such as the proprietary web applications Google Maps and Bing Maps, as well as the free and open-source alternative OpenStreetMap. These services give the public access to huge amounts of geographic data.

Some of them, like Google Maps and OpenLayers, expose an API that enable users to create custom applications. These toolkits commonly offer street maps, aerial/ satellite imagery, geocoding, searches, and routing functionality. Web mapping has also uncovered the potential of crowdsourcing geodata in projects like OpenStreetMap, which is a collaborative project to create a free editable map of the world.

Adding the Dimension of Time

The condition of the Earth's surface, atmosphere, and subsurface can be examined by feeding satellite data into a GIS. GIS technology gives researchers the ability to examine the variations in Earth processes over days, months, and years. As an example, the changes in vegetation vigor through a growing season can be animated to determine when drought was most extensive in a particular region. The resulting graphic represents a rough measure of plant health. Working with two variables over time would then allow researchers to detect regional differences in the lag between a decline in rainfall and its effect on vegetation.

GIS technology and the availability of digital data on regional and global scales enable such analyses. The satellite sensor output used to generate a vegetation graphic is produced for example by the Advanced Very High Resolution Radiometer (AVHRR). This sensor system detects the amounts of energy reflected from the Earth's surface across various bands of the spectrum for surface areas of about 1 square kilometer. The satellite sensor produces images of a particular location on the Earth twice a day. AVHRR and more recently the Moderate-Resolution Imaging Spectroradiometer (MODIS) are only two of many sensor systems used for Earth surface analysis. More sensors will follow, generating ever greater amounts of data.

In addition to the integration of time in environmental studies, GIS is also being explored for its ability to track and model the progress of humans throughout their daily routines. A concrete example of progress in this area is the recent release of time-specific population data by the U.S. Census. In this data set, the populations of cities are shown for daytime and evening hours highlighting the pattern of concentration and dispersion generated by North American commuting patterns. The manipulation and generation of data required to produce this data would not have been possible without GIS.

Using models to project the data held by a GIS forward in time have enabled planners to test policy decisions using spatial decision support systems.

Semantics

Tools and technologies emerging from the W3C's Data Activity are proving useful for data integration problems in information systems. Correspondingly, such technologies have been proposed as a means to facilitate interoperability and data reuse among GIS applications. and also to enable new analysis mechanisms.

Ontologies are a key component of this semantic approach as they allow a formal, machine-readable specification of the concepts and relationships in a given domain. This in turn allows a GIS to focus on the intended meaning of data rather than its syntax or structure. For example, reasoning that a land cover type classified as deciduous needle leaf trees in one dataset is a specialization or subset of land cover type forest in another more roughly classified dataset can help a GIS automatically merge the two datasets under the more general land cover classification. Tentative ontologies have been developed in areas related to GIS applications, for example the hydrology ontology developed by the Ordnance Survey in the United Kingdom and the SWEET ontologies developed by NASA's Jet Propulsion Laboratory. Also, simpler ontologies and semantic metadata standards are being proposed by the W3C Geo Incubator Group to represent geospatial data on the web. GeoSPARQL is a standard developed by the Ordnance Survey, United States Geological Survey, Natural Resources Canada, Australia's Commonwealth Scientific and Industrial Research Organization and others to support ontology creation and reasoning using well-understood OGC literals (GML, WKT), topological relationships (Simple Features, RCC8, DE-9IM), RDF and the SPARQL database query protocols.

Recent research results in this area can be seen in the International Conference on Geospatial Semantics and the Terra Cognita – Directions to the Geospatial Semantic Web workshop at the International Semantic Web Conference.

Implications of GIS in Society

With the popularization of GIS in decision making, scholars have begun to scrutinize the social and political implications of GIS. GIS can also be misused to distort reality for individual and political gain. It has been argued that the production, distribution, utilization, and representation of geographic information are largely related with the social context and has the potential to increase citizen trust in government. Other related topics include discussion on copyright, privacy, and censorship. A more optimistic social approach to GIS adoption is to use it as a tool for public participation.

GIS FILE FORMATS

A GIS file format is a standard of encoding geographical information into a computer file. They are created mainly by government mapping agencies (such as the USGS or National Geospatial-Intelligence Agency) or by GIS software developers.

Raster

A raster data type is, in essence, any type of digital image represented by reducible and enlargeable grids. Anyone who is familiar with digital photography will recognize the Raster graphics pixel as the smallest individual grid unit building block of an image, usually not readily identified as an artifact shape until an image is produced on a very large scale. A combination of the pixels making up an image color formation scheme will compose details of an image, as is distinct from the commonly used points, lines, and polygon area location symbols of scalable vector graphics as the basis of the vector model of area attribute rendering. While a digital image is concerned with its output blending together its grid based details as an identifiable representation of reality, in a photograph or art image transferred into a computer, the raster data type will reflect a digitized abstraction of reality dealt with by grid populating tones or objects, quantities, cojoined or open boundaries, and map relief schemas. Aerial photos are one commonly used form of raster data, with one primary purpose in mind: to display a detailed image on a map area, or for the purposes of rendering its identifiable objects by digitization. Additional raster data sets used by a GIS will contain information regarding elevation, a digital elevation model, or reflectance of a particular wavelength of light, Landsat, or other electromagnetic spectrum indicators.

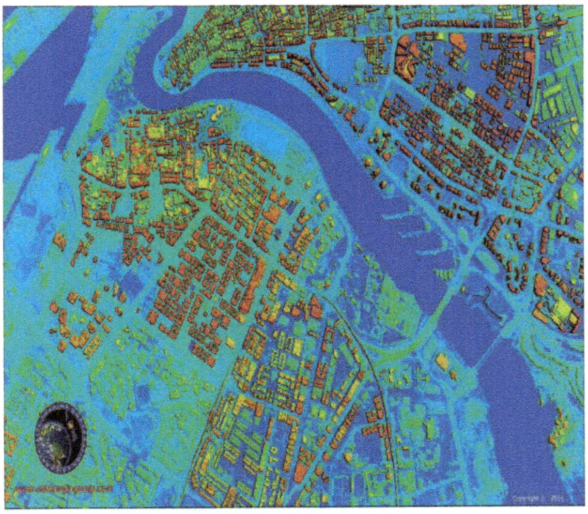

Digital elevation model, map (image), and vector data

Raster data type consists of rows and columns of cells, with each cell storing a single value. Raster data can be images raster images) with each pixel (or cell) containing a color value. Additional values recorded for each cell may be a discrete value, such as land use, a continuous value, such as temperature, or a null value if no data is available. While a raster cell stores a single value, it can be extended by using raster bands to represent RGB (red, green, blue) colors, colormaps (a mapping between a thematic code and RGB value), or an extended attribute table with one row for each unique cell value. The resolution of the raster data set is its cell width in ground units.

Raster data is stored in various formats; from a standard file-based structure of TIFF, JPEG, etc. to binary large object (BLOB) data stored directly in a relational database management system (RDBMS) similar to other vector-based feature classes. Database storage, when properly indexed, typically allows for quicker retrieval of the raster data but can require storage of millions of significantly sized records.

Vector

In a GIS, geographical features are often expressed as vectors, by considering those features as geometrical shapes. Different geographical features are expressed by different types of geometry:

- Points

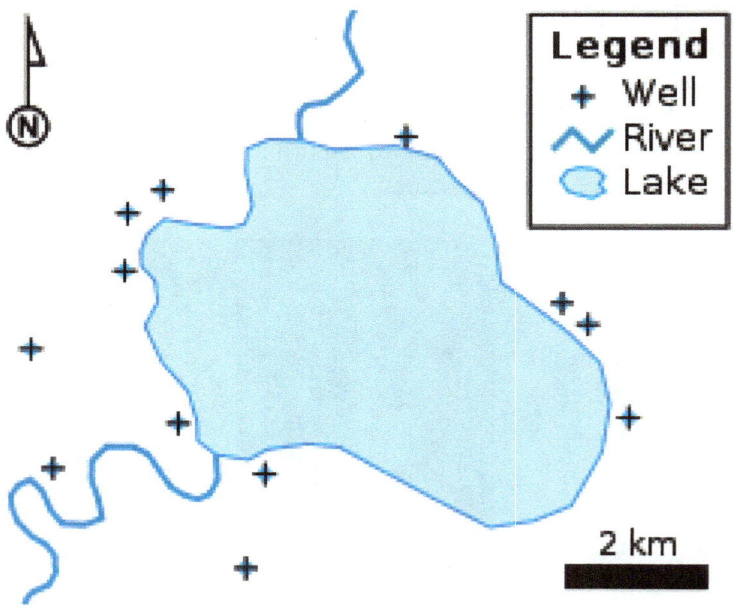

A simple vector map, using each of the vector elements: points for wells, lines for rivers, and a polygon for the lake

Zero-dimensional points are used for geographical features that can best be expressed by a single point reference—in other words, by simple location. Examples include wells, peaks, features of interest, and trailheads. Points convey the least amount of information of these file types. Points can also be used to represent areas when displayed at a small scale. For example, cities on a map of the world might be represented by points rather than polygons. No measurements are possible with point features.

- Lines or polylines

One-dimensional lines or polylines are used for linear features such as rivers, roads, railroads, trails, and topographic lines. Again, as with point features, linear features displayed at a small scale will be represented as linear features rather than as a polygon. Line features can measure distance.

- Polygons

Two-dimensional polygons are used for geographical features that cover a particular area of the earth's surface. Such features may include lakes, park boundaries, buildings, city boundaries, or land uses. Polygons convey the most amount of information of the file types. Polygon features can measure perimeter and area.

Each of these geometries are linked to a row in a database that describes their attributes. For example, a database that describes lakes may contain a lake's depth, water quality, pollution level. This information can be used to make a map to describe a particular attribute of the dataset. For example, lakes could be colored depending on level of pollution. Different geometries can also be compared. For example, the GIS could be used to identify all wells (point geometry) that are within one kilometer of a lake (polygon geometry) that has a high level of pollution.

Vector features can be made to respect spatial integrity through the application of topology rules such as 'polygons must not overlap'. Vector data can also be used to represent continuously varying phenomena. Contour lines and triangulated irregular networks (TIN) are used to represent elevation or other continuously changing values. TINs record values at point locations, which are connected by lines to form an irregular mesh of triangles. The face of the triangles represent the terrain surface.

Advantages and Disadvantages

There are some important advantages and disadvantages to using a raster or vector data model to represent reality:

- Raster datasets record a value for all points in the area covered which may require more storage space than representing data in a vector format that can store data only where needed.
- Raster data is computationally less expensive to render than vector graphics
- There are transparency and aliasing problems when overlaying multiple stacked pieces of raster images
- Vector data allows for visually smooth and easy implementation of overlay operations, especially in terms of graphics and shape-driven information like maps, routes and custom fonts, which are more difficult with raster data.
- Vector data can be displayed as vector graphics used on traditional maps, whereas raster data will appear as an image that may have a blocky appearance for object boundaries. (depending on the resolution of the raster file)
- Vector data can be easier to register, scale, and re-project, which can simplify combining vector layers from different sources.

- Vector data is more compatible with relational database environments, where they can be part of a relational table as a normal column and processed using a multitude of operators.
- Vector file sizes are usually smaller than raster data, which can be tens, hundreds or more times larger than vector data (depending on resolution).
- Vector data is simpler to update and maintain, whereas a raster image will have to be completely reproduced. (Example: a new road is added).
- Vector data allows much more analysis capability, especially for "networks" such as roads, power, rail, telecommunications, etc. (Examples: Best route, largest port, airfields connected to two-lane highways). Raster data will not have all the characteristics of the features it displays.

Non-Spatial Data

Additional non-spatial data can also be stored along with the spatial data represented by the coordinates of a vector geometry or the position of a raster cell. In vector data, the additional data contains attributes of the feature. For example, a forest inventory polygon may also have an identifier value and information about tree species. In raster data the cell value can store attribute information, but it can also be used as an identifier that can relate to records in another table.

Software is currently being developed to support spatial and non-spatial decision-making, with the solutions to spatial problems being integrated with solutions to non-spatial problems. The end result with these flexible spatial decision-making support systems (FSDSSs) is expected to be that non-experts will be able to use GIS, along with spatial criteria, and simply integrate their non-spatial criteria to view solutions to multi-criteria problems. This system is intended to assist decision-making.

MAP PROJECTION

Commonly, a map projection is a systematic transformation of the latitudes and longitudes of locations on the surface of a sphere or an ellipsoid into locations on a plane. Map projections are necessary for creating maps. All map projections distort the surface in some fashion. Depending on the purpose of the map, some distortions are acceptable and others are not; therefore, different map projections exist in order to preserve some properties of the sphere-like body at the expense of other properties. There is no limit to the number of possible map projections.

More generally, the surfaces of planetary bodies can be mapped even if they are too irregular to be modeled well with a sphere or ellipsoid; see below. Even more generally, projections are the subject of several pure mathematical fields, including differential geometry and projective geometry. However, "map projection" refers specifically to a cartographic projection.

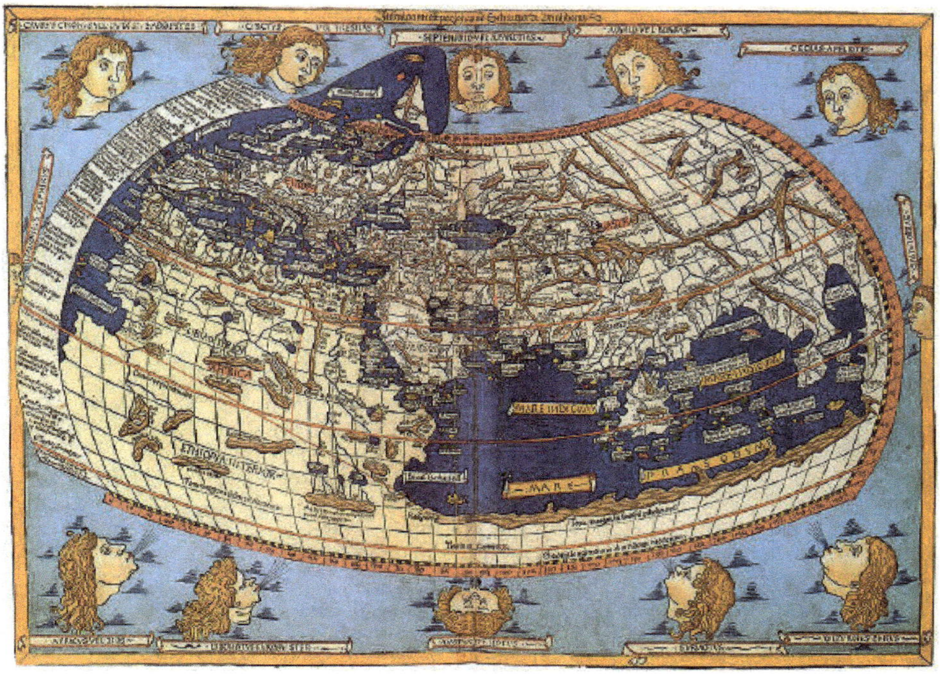

A medieval depiction of the Ecumene (1482, Johannes Schnitzer, engraver), constructed after the coordinates in Ptolemy's Geography and using his second map projection.

Background

Maps can be more useful than globes in many situations: they are more compact and easier to store; they readily accommodate an enormous range of scales; they are viewed easily on computer displays; they can facilitate measuring properties of the terrain being mapped; they can show larger portions of the Earth's surface at once; and they are cheaper to produce and transport. These useful traits of maps motivate the development of map projections.

However, Carl Friedrich Gauss's Theorema Egregium proved that a sphere's surface cannot be represented on a plane without distortion. The same applies to other reference surfaces used as models for the Earth. Since any map projection is a representation of one of those surfaces on a plane, all map projections distort. Every distinct map projection distorts in a distinct way. The study of map projections is the characterization of these distortions.

Projection is not limited to perspective projections, such as those resulting from casting a shadow on a screen, or the rectilinear image produced by a pinhole camera on a flat film plate. Rather, any mathematical function transforming coordinates from the curved surface to the plane is a projection. Few projections in actual use are perspective.

For simplicity, most of this chapter assumes that the surface to be mapped is that of a sphere. In reality, the Earth and other large celestial bodies are generally better modeled as oblate spheroids, whereas small objects such as asteroids often have irregular shapes. These other surfaces can be mapped as well. Therefore, more generally, a map projection is any method of "flattening" into a plane a continuous curved surface.

Metric Properties of Maps

Many properties can be measured on the Earth's surface independent of its geography. Some of these properties are:

- Area
- Shape
- Direction
- Bearing
- Distance
- Scale

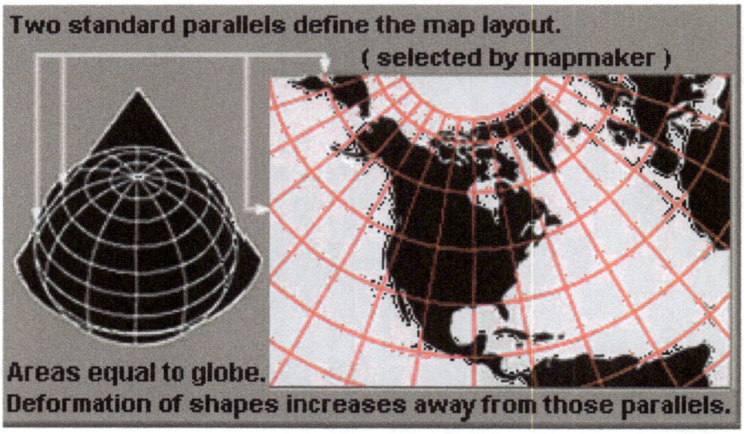

An Albers projection shows areas accurately, but distorts shapes.

Map projections can be constructed to preserve at least one of these properties, though only in a limited way for most. Each projection preserves or compromises or approximates basic metric properties in different ways. The purpose of the map determines which projection should form the base for the map. Because many purposes exist for maps, many projections have been created to suit those purposes.

Another consideration in the configuration of a projection is its compatibility with data sets to be used on the map. Data sets are geographic information; their collection depends on the chosen datum (model) of the Earth. Different datums assign slightly different coordinates to the same location, so in large scale maps, such as those from national mapping systems, it is important to match the datum to the projection. The slight differences in coordinate assignation between different datums is not a concern for world maps or other vast territories, where such differences get shrunk to imperceptibility.

Which Projection is Best?

The mathematics of projection do not permit any particular map projection to be "best" for everything. Something will always get distorted. Therefore, a diversity of projections exists to service the many uses of maps and their vast range of scales.

Modern national mapping systems typically employ a transverse Mercator or close variant for large-scale maps in order to preserve conformality and low variation in scale over small areas. For smaller-scale maps, such as those spanning continents or the entire world, many projections are in common use according to their fitness for the purpose.

Thematic maps normally require an equal area projection so that phenomena per unit area are shown in correct proportion. However, representing area ratios correctly necessarily distorts shapes more than many maps that are not equal-area. Hence reference maps of the world often appear on compromise projections instead. Due to distortions inherent in any map of the world, the choice of projection becomes largely one of aesthetics.

The Mercator projection, developed for navigational purposes, has often been used in world maps where other projections would have been more appropriate. This problem has long been recognized even outside professional circles. For example, a 1943 New York Times editorial states:

The time has come to discard [the Mercator] for something that represents the continents and directions less deceptively... Although its usage... has diminished... it is still highly popular as a wall map apparently in part because, as a rectangular map, it fills a rectangular wall space with more map, and clearly because its familiarity breeds more popularity.

A controversy in the 1980s over the Peters map motivated the American Cartographic Association (now Cartography and Geographic Information Society) to produce a series of booklets (including Which Map Is Best) designed to educate the public about map projections and distortion in maps. In 1989 and 1990, after some internal debate, seven North American geographic organizations adopted a resolution recommending against using any rectangular projection (including Mercator and Gall–Peters) for reference maps of the world.

Distortion

The classical way of showing the distortion inherent in a projection is to use Tissot's indicatrix. For a given point, using the scale factor h along the meridian, the scale factor k along the parallel, and the angle θ' between them, Nicolas Tissot described how to construct an ellipse that characterizes the amount and orientation of the components of distortion.

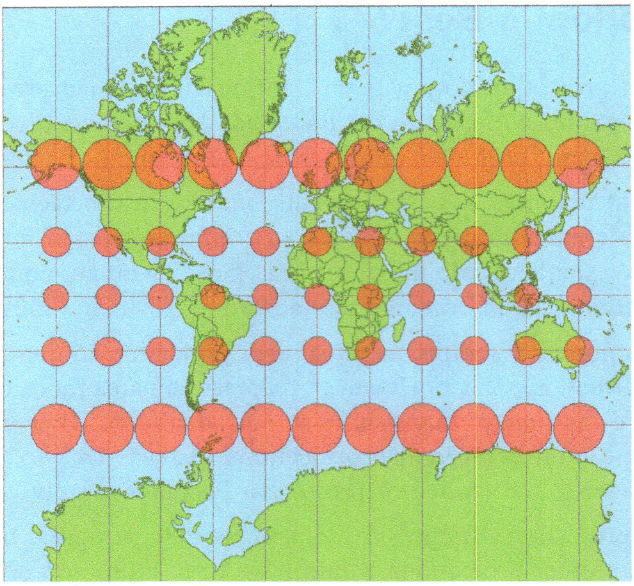

Tissot's Indicatrices on the Mercator projection

By spacing the ellipses regularly along the meridians and parallels, the network of indicatrices shows how distortion varies across the map.

Construction of a Map Projection

The creation of a map projection involves two steps:

- Selection of a model for the shape of the Earth or planetary body (usually choosing between a sphere or ellipsoid). Because the Earth's actual shape is irregular, information is lost in this step.
- Transformation of geographic coordinates (longitude and latitude) to Cartesian (x,y) or polar plane coordinates. Cartesian coordinates normally have a simple relation to eastings and northings defined on a grid superimposed on the projection.

Some of the simplest map projections are literal projections, as obtained by placing a light source at some definite point relative to the globe and projecting its features onto a specified surface. This is not the case for most projections, which are defined **only** in terms of mathematical formulae that have no direct geometric interpretation.

Choosing a Projection Surface

A surface that can be unfolded or unrolled into a plane or sheet without stretching, tearing or shrinking is called a developable surface. The cylinder, cone and the plane are all developable surfaces. The sphere and ellipsoid do not have developable surfaces, so any projection of them onto a plane will have to distort the image. (To compare, one cannot flatten an orange peel without tearing and warping it.)

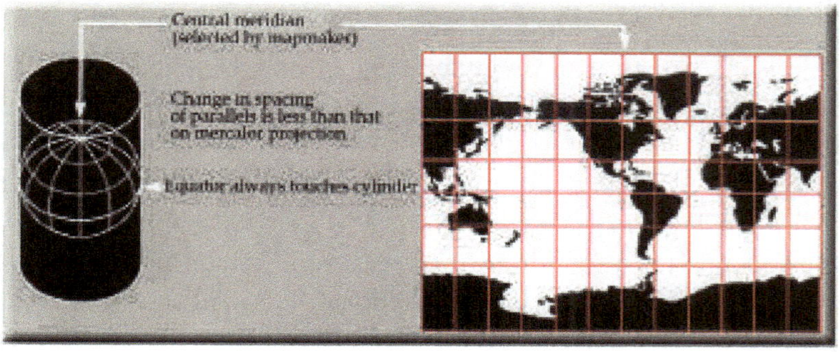

A Miller cylindrical projection maps the globe onto a cylinder.

One way of describing a projection is first to project from the Earth's surface to a developable surface such as a cylinder or cone, and then to unroll the surface into a plane. While the first step inevitably distorts some properties of the globe, the developable surface can then be unfolded without further distortion.

Aspect of the Projection

Once a choice is made between projecting onto a cylinder, cone, or plane, the **aspect** of the shape must be specified. The aspect describes how the developable surface is placed relative to the globe: it may be normal (such that the surface's axis of symmetry coincides with the Earth's axis), transverse (at right angles to the Earth's axis) or oblique (any angle in between).

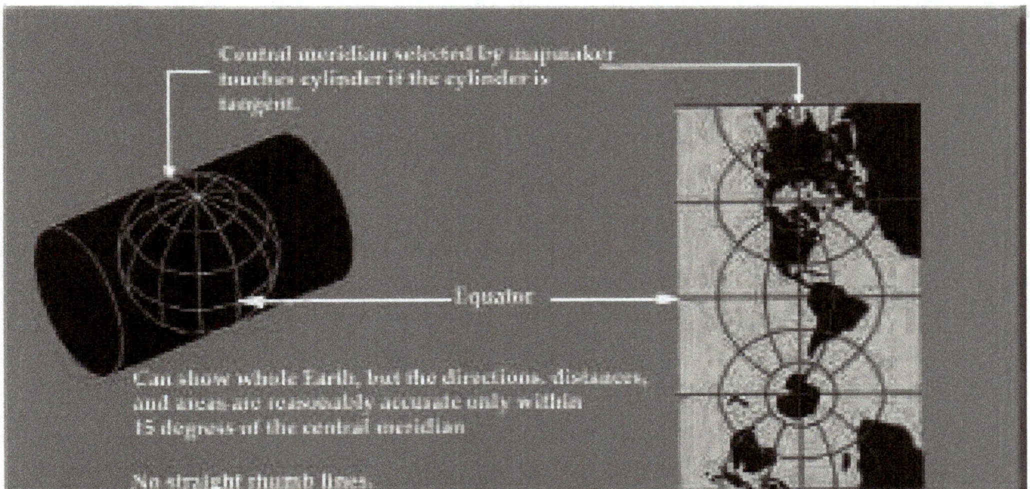

This transverse Mercator projection is mathematically the same as a standard Mercator, but oriented around a different axis.

Notable Lines

The developable surface may also be either tangent or secant to the sphere or ellipsoid. Tangent means the surface touches but does not slice through the globe; secant means the surface does slice through the globe. Moving the developable surface away from contact with the globe never preserves or optimizes metric properties, so that possibility is not discussed further here.

Tangent and secant lines (standard lines) are represented undistorted. If these lines are a parallel of latitude, as in conical projections, it is called a standard parallel. The central meridian is the meridian to which the globe is rotated before projecting. The central meridian (usually written λ_0) and a parallel of origin (usually written φ_0) are often used to define the origin of the map projection.

Scale

A globe is the only way to represent the earth with constant scale throughout the entire map in all directions. A map cannot achieve that property for any area, no matter how small. It can, however, achieve constant scale along specific lines.

Some possible properties are:

- The scale depends on location, but not on direction. This is equivalent to preservation of angles, the defining characteristic of a conformal map.
- Scale is constant along any parallel in the direction of the parallel. This applies for any cylindrical or pseudocylindrical projection in normal aspect.
- Combination of the above: the scale depends on latitude only, not on longitude or direction. This applies for the Mercator projection in normal aspect.
- Scale is constant along all straight lines radiating from a particular geographic location. This is the defining characteristic of an equidistant projection such as the Azimuthal equidistant projection. There are also projections (Maurer, Close) where true distances from two points are preserved.[:234]

Choosing a Model for the Shape of the Body

Projection construction is also affected by how the shape of the Earth or planetary body is approximated. In the following section on projection categories, the earth is taken as a sphere in order to simplify the discussion. However, the Earth's actual shape is closer to an oblate ellipsoid. Whether spherical or ellipsoidal, the principles discussed hold without loss of generality.

Selecting a model for a shape of the Earth involves choosing between the advantages and disadvantages of a sphere versus an ellipsoid. Spherical models are useful for small-scale maps such as world atlases and globes, since the error at that scale is not usually noticeable or important enough to justify using the more complicated ellipsoid. The ellipsoidal model is commonly used to construct topographic maps and for other large- and medium-scale maps that need to accurately depict the land surface. Auxiliary latitudes are often employed in projecting the ellipsoid.

A third model is the geoid, a more complex and accurate representation of Earth's shape coincident with what mean sea level would be if there were no winds, tides, or land. Compared to the best fitting ellipsoid, a geoidal model would change the characterization of important properties such as distance, conformality and equivalence. Therefore, in geoidal projections that preserve such properties, the mapped graticule would deviate from a mapped ellipsoid's graticule. Normally the geoid is not used as an Earth model for projections, however, because Earth's shape is very regular, with the undulation of the geoid amounting to less than 100 m from the ellipsoidal model out of the 6.3 million m Earth radius. For irregular planetary bodies such as asteroids, however, sometimes models analogous to the geoid are used to project maps from.

Classification

A fundamental projection classification is based on the type of projection surface onto which the globe is conceptually projected. The projections are described in terms of placing a gigantic surface in contact with the earth, followed by an implied scaling operation. These surfaces are cylindrical (e.g. Mercator), conic (e.g. Albers), or azimuthal or plane (e.g. stereographic). Many mathematical projections, however, do not neatly fit into any of these three conceptual projection methods. Hence other peer categories have been described in the literature, such as pseudoconic, pseudocylindrical, pseudoazimuthal, retroazimuthal, and polyconic.

Another way to classify projections is according to properties of the model they preserve. Some of the more common categories are:

- Preserving direction (azimuthal or zenithal), a trait possible only from one or two points to every other point
- Preserving shape locally (conformal or orthomorphic)
- Preserving area (equal-area or equiareal or equivalent or authalic)
- Preserving distance (equidistant), a trait possible only between one or two points and every other point
- Preserving shortest route, a trait preserved only by the gnomonic projection

Because the sphere is not a developable surface, it is impossible to construct a map projection that is both equal-area and conformal.

Projections by Surface

The three developable surfaces (plane, cylinder, cone) provide useful models for understanding, describing, and developing map projections. However, these models are limited in two fundamental ways. For one thing, most world projections in actual use do not fall into any of those categories. For another thing, even most projections that do fall into those categories are not naturally attainable through physical projection. As L.P. Lee notes,

No reference has been made in the above definitions to cylinders, cones or planes. The projections are termed cylindric or conic because they can be regarded as developed on a cylinder or a cone, as the case may be, but it is as well to dispense with picturing cylinders and cones, since they have given rise to much misunderstanding. Particularly is this so with regard to the conic projections with two standard parallels: they may be regarded as developed on cones, but they are cones which bear no simple relationship to the sphere. In reality, cylinders and cones provide us with convenient descriptive terms, but little else.

Lee's objection refers to the way the terms cylindrical, conic, and planar (azimuthal) have been abstracted in the field of map projections. If maps were projected as in light shining through a globe onto a developable surface, then the spacing of parallels would follow a very limited set of possibilities. Such a cylindrical projection (for example) is one which:

- Is rectangular;
- Has straight vertical meridians, spaced evenly;
- Has straight parallels symmetrically placed about the equator;
- Has parallels constrained to where they fall when light shines through the globe onto the cylinder, with the light source someplace along the line formed by the intersection of the prime meridian with the equator, and the center of the sphere.

(If you rotate the globe before projecting then the parallels and meridians will not necessarily still be straight lines. Rotations are normally ignored for the purpose of classification.)

Where the light source emanates along the line described in this last constraint is what yields the differences between the various "natural" cylindrical projections. But the term cylindrical as used in the field of map projections relaxes the last constraint entirely. Instead the parallels can be placed according to any algorithm the designer has decided suits the needs of the map. The famous Mercator projection is one in which the placement of parallels does not arise by "projection"; instead parallels are placed how they need to be in order to satisfy the property that a course of constant bearing is always plotted as a straight line.

Cylindrical

The term "normal cylindrical projection" is used to refer to any projection in which meridians are mapped to equally spaced vertical lines and circles of latitude (parallels) are mapped to horizontal lines.

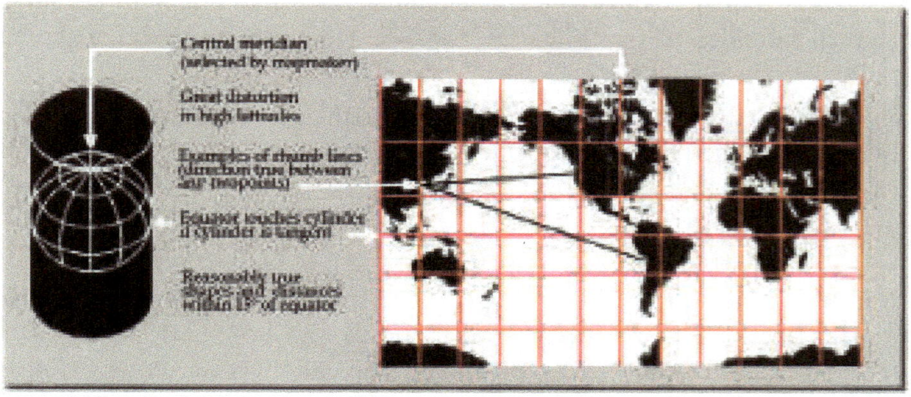

The Mercator projection shows rhumbs as straight lines. A rhumb is a course of constant bearing. Bearing is the compass direction of movement.

The mapping of meridians to vertical lines can be visualized by imagining a cylinder whose axis coincides with the Earth's axis of rotation. This cylinder is wrapped around the Earth, projected onto, and then unrolled.

By the geometry of their construction, cylindrical projections stretch distances east-west. The amount of stretch is the same at any chosen latitude on all cylindrical projections, and is given by the secant of the latitude as a multiple of the equator's scale. The various cylindrical projections are distinguished from each other solely by their north-south stretching (where latitude is given by φ):

- North-south stretching equals east-west stretching (sec φ): The east-west scale matches the north-south scale: conformal cylindrical or Mercator; this distorts areas excessively in high latitudes.
- North-south stretching grows with latitude faster than east-west stretching ($\sec^2 \varphi$): The cylindric perspective (or central cylindrical) projection; unsuitable because distortion is even worse than in the Mercator projection.
- North-south stretching grows with latitude, but less quickly than the east-west stretching: such as the Miller cylindrical projection (sec 4φ/5).
- North-south distances neither stretched nor compressed (1): equirectangular projection or plate carrée.
- North-south compression equals the cosine of the latitude (the reciprocal of east-west stretching): equal-area cylindrical. This projection has many named specializations differing only in the scaling constant, such as the Gall–Peters or Gall orthographic (undistorted at the 45° parallels), Behrmann (undistorted at the 30° parallels), and Lambert cylindrical equal-area (undistorted at the equator). Since this projection scales north-south distances by the reciprocal of east-west stretching, it preserves area at the expense of shapes.

In the first case (Mercator), the east-west scale always equals the north-south scale. In the second case (central cylindrical), the north-south scale exceeds the east-west scale everywhere away from the equator. Each remaining case has a pair of secant lines—a pair of identical latitudes of opposite sign (or else the equator) at which the east-west scale matches the north-south-scale.

Normal cylindrical projections map the whole Earth as a finite rectangle, except in the first two cases, where the rectangle stretches infinitely tall while retaining constant width.

Pseudocylindrical

Pseudocylindrical projections represent the central meridian as a straight line segment. Other meridians are longer than the central meridian and bow outward away from the central meridian. Pseudocylindrical projections map parallels as straight lines. Along parallels, each point from the surface is mapped at a distance from the central meridian that is proportional to its difference in longitude from the central meridian.

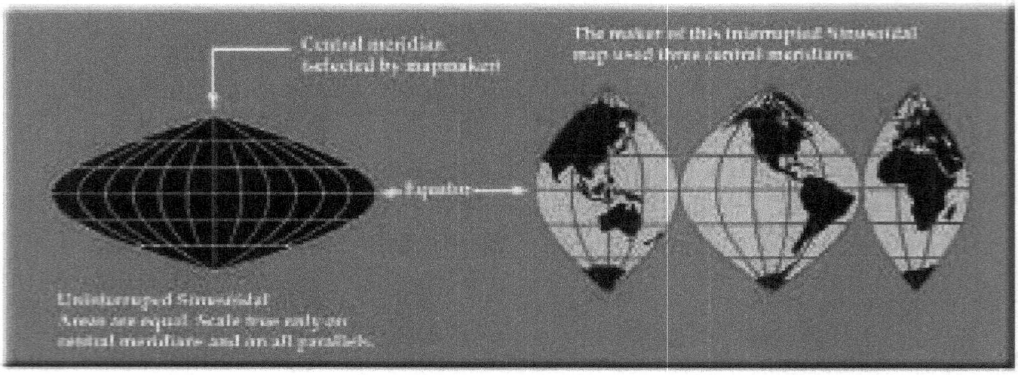

A sinusoidal projection shows relative sizes accurately, but grossly distorts shapes. Distortion can be reduced by "interrupting" the map.

On a pseudocylindrical map, any point further from the equator than some other point has a higher latitude than the other point, preserving north-south relationships. This trait is useful when illustrating phenomena that depend on latitude, such as climate. Examples of pseudocylindrical projections include:

- Sinusoidal, which was the first pseudocylindrical projection developed. Vertical scale and horizontal scale are the same throughout, resulting in an equal-area map. On the map, as in reality, the length of each parallel is proportional to the cosine of the latitude. Thus the shape of the map for the whole earth is the region between two symmetric rotated cosine curves. The true distance between two points on the same meridian corresponds to the distance on the map between the two parallels, which is smaller than the distance between the two points on the map. The distance between two points on the same parallel is true. The area of any region is true.
- Collignon projection, which in its most common forms represents each meridian as two straight line segments, one from each pole to the equator.

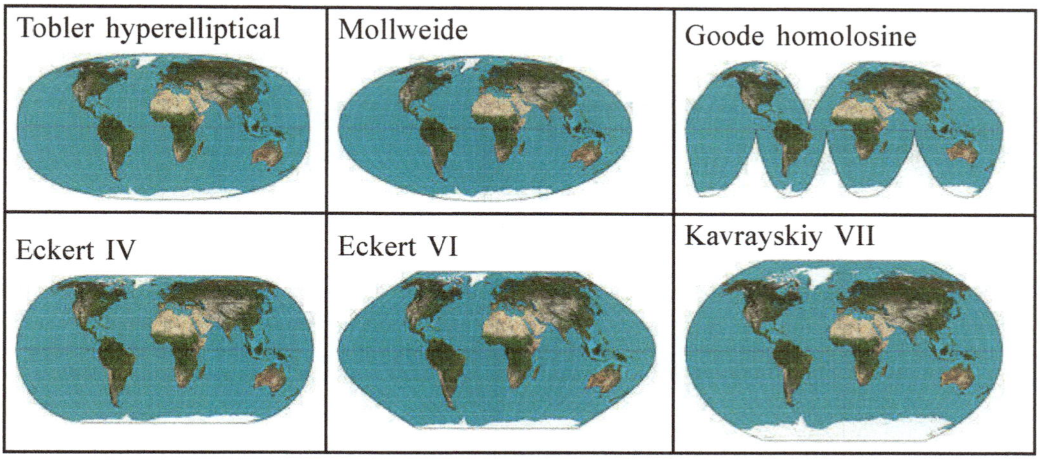

Hybrid

The HEALPix projection combines an equal-area cylindrical projection in equatorial regions with the Collignon projection in polar areas.

Conic

The term "conic projection" is used to refer to any projection in which meridians are mapped to equally spaced lines radiating out from the apex and circles of latitude (parallels) are mapped to circular arcs centered on the apex.

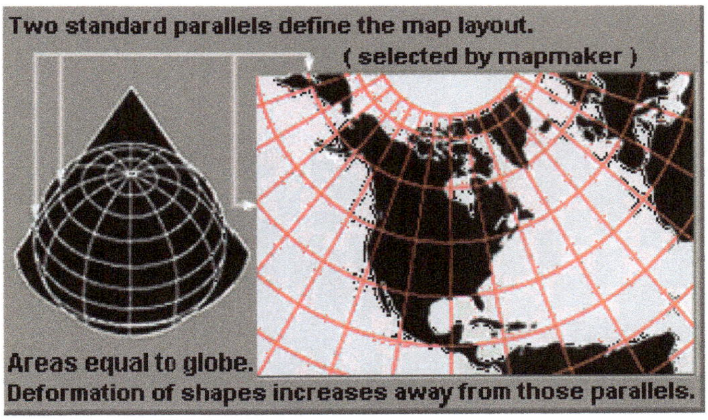

Albers conic.

When making a conic map, the map maker arbitrarily picks two standard parallels. Those standard parallels may be visualized as secant lines where the cone intersects the globe—or, if the map maker chooses the same parallel twice, as the tangent line where the cone is tangent to the globe. The resulting conic map has low distortion in scale, shape, and area near those standard parallels. Distances along the parallels to the north of both standard parallels or to the south of both standard parallels are stretched; distances along parallels between the standard parallels are compressed. When a single standard parallel is used, distances along all other parallels are stretched.

The most popular conic maps include:

- Equidistant conic, which keeps parallels evenly spaced along the meridians to preserve a constant distance scale along each meridian, typically the same or similar scale as along the standard parallels.
- Albers conic, which adjusts the north-south distance between non-standard parallels to compensate for the east-west stretching or compression, giving an equal-area map.
- Lambert conformal conic, which adjusts the north-south distance between non-standard parallels to equal the east-west stretching, giving a conformal map.

Pseudoconic

- Bonne
- Werner cordiform, upon which distances are correct from one pole, as well as along all parallels.
- Continuous American polyconic

Azimuthal (Projections onto a Plane)

Azimuthal projections have the property that directions from a central point are preserved and therefore great circles through the central point are represented by straight lines on the map. Usually these projections also have radial symmetry in the scales and hence in the distortions: map distances from the central point are computed by a function r(d) of the true distance d, independent of the angle; correspondingly, circles with the central point as center are mapped into circles which have as center the central point on the map.

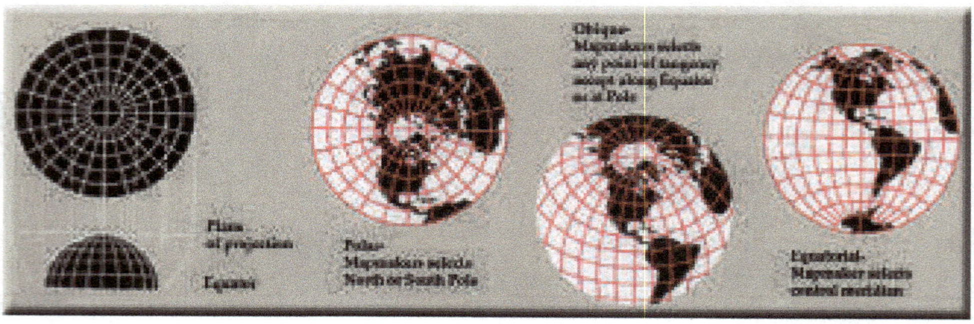

An azimuthal equidistant projection shows distances and directions accurately from the center point, but distorts shapes and sizes elsewhere.

The mapping of radial lines can be visualized by imagining a plane tangent to the Earth, with the central point as tangent point.

The radial scale is $r'(d)$ and the transverse scale $r(d)/(R \sin d/R)$ where R is the radius of the Earth.

Some azimuthal projections are true perspective projections; that is, they can be constructed mechanically, projecting the surface of the Earth by extending lines from a point of perspective (along an infinite line through the tangent point and the tangent point's antipode) onto the plane:

- The gnomonic projection displays great circles as straight lines. Can be constructed by using a point of perspective at the center of the Earth. r(d) = c tan d/R; so that even just a hemisphere is already infinite in extent.
- The General Perspective projection can be constructed by using a point of perspective outside the earth. Photographs of Earth (such as those from the International Space Station) give this perspective.
- The orthographic projection maps each point on the earth to the closest point on the plane. Can be constructed from a point of perspective an infinite distance from the tangent point; r(d) = c sin d/R. Can display up to a hemisphere on a finite circle. Photographs of Earth from far enough away, such as the Moon, give this perspective.
- The azimuthal conformal projection, also known as the stereographic projection, can be constructed by using the tangent point's antipode as the point of perspective. r(d) = c tan d/2R; the scale is c/(2R $\cos^2$ d/2R). Can display nearly the entire sphere's surface on a finite circle. The sphere's full surface requires an infinite map.

Other azimuthal projections are not true perspective projections:

- Azimuthal equidistant: r(d) = cd; it is used by amateur radio operators to know the direction to point their antennas toward a point and see the distance to it. Distance from the tangent point on the map is proportional to surface distance on the earth (; for the case where the tangent point is the North Pole, see the flag of the United Nations)
- Lambert azimuthal equal-area. Distance from the tangent point on the map is proportional to straight-line distance through the earth: r(d) = c sin d/2R
- Logarithmic azimuthal is constructed so that each point's distance from the center of the map is the logarithm of its distance from the tangent point on the Earth. r(d) = c ln d/d_0); locations closer than at a distance equal to the constant d_0 are not shown.

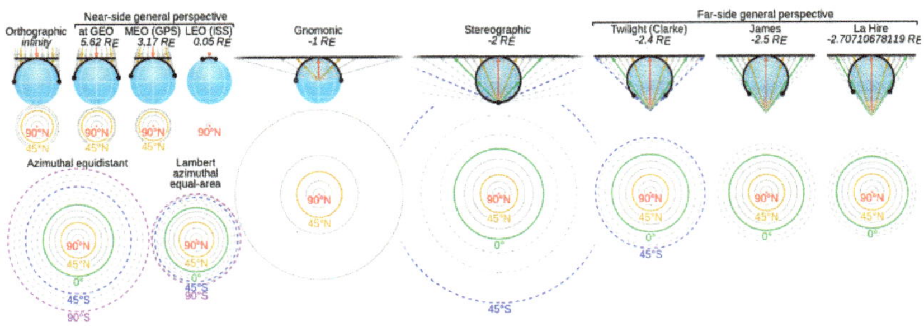

Comparison of some azimuthal projections centred on 90° N at the same scale, ordered by projection altitude in Earth radii.

Projections by Preservation of a Metric Property

Conformal

Conformal, or orthomorphic, map projections preserve angles locally, implying that they map infinitesimal circles of constant size anywhere on the Earth to infinitesimal circles of varying sizes on the map. In contrast, mappings that are not conformal distort most such small circles into ellipses of distortion.

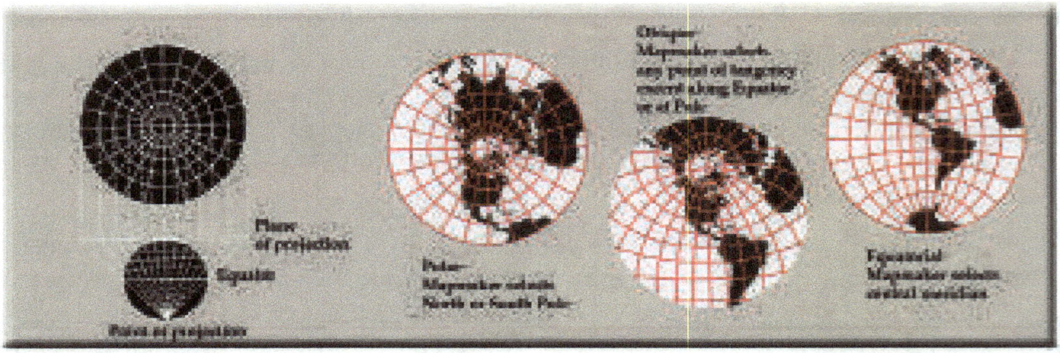

A stereographic projection is conformal and perspective but not equal area or equidistant.

An important consequence of conformality is that relative angles at each point of the map are correct, and the local scale (although varying throughout the map) in every direction around any one point is constant. These are some conformal projections:

- Mercator: Rhumb lines are represented by straight segments
- Transverse Mercator
- Stereographic: Any circle of a sphere, great and small, maps to a circle or straight line.
- Roussilhe
- Lambert conformal conic
- Peirce quincuncial projection
- Adams hemisphere-in-a-square projection
- Guyou hemisphere-in-a-square projection

Equal-Area

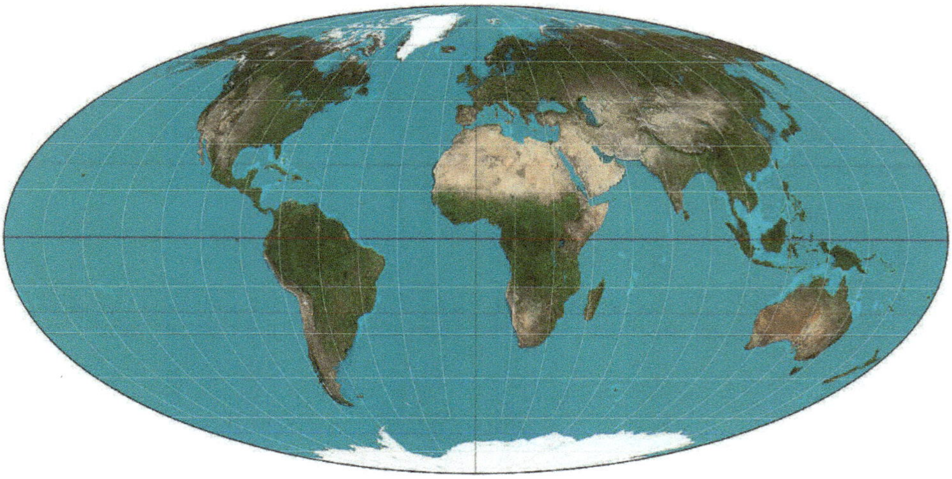

The equal-area Mollweide projection

Equal-area maps preserve area measure, generally distorting shapes in order to do that. Equal-area maps are also called equivalent or authalic. These are some projections that preserve area:

- Albers conic
- Bonne
- Bottomley
- Collignon
- Cylindrical equal-area
- Eckert II, IV and VI
- Gall orthographic (also known as Gall–Peters, or Peters, projection)
- Goode's homolosine
- Hammer
- Hobo–Dyer
- Lambert azimuthal equal-area
- Lambert cylindrical equal-area
- Mollweide
- Sinusoidal
- Snyder's equal-area polyhedral projection, used for geodesic grids.
- Tobler hyperelliptical
- Werner

Equidistant

These are some projections that preserve distance from some standard point or line:

- Equirectangular—distances along meridians are conserved
- Plate carrée—an Equirectangular projection centered at the equator
- Azimuthal equidistant—distances along great circles radiating from centre are conserved
- Equidistant conic
- Sinusoidal—distances along parallels are conserved
- Werner cordiform distances from the North Pole are correct as are the curved distance on parallels
- Soldner
- Two-point equidistant: two "control points" are arbitrarily chosen by the map maker. Distance from any point on the map to each control point is proportional to surface distance on the earth.

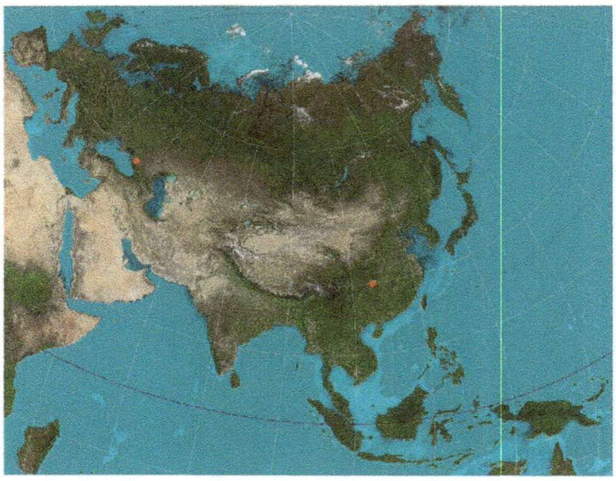

A two-point equidistant projection of Eurasia

Gnomonic

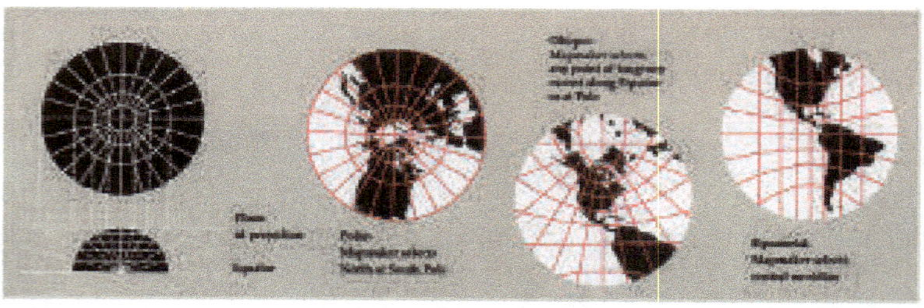

The Gnomonic projection is thought to be the oldest map projection, developed by Thales in the 6th century BC

Great circles are displayed as straight lines:

- Gnomonic projection

Retroazimuthal

Direction to a fixed location B (the bearing at the starting location A of the shortest route) corresponds to the direction on the map from A to B:

- Littrow—the only conformal retroazimuthal projection
- Hammer retroazimuthal—also preserves distance from the central point
- Craig retroazimuthal aka Mecca or Qibla—also has vertical meridians

Compromise Projections

Compromise projections give up the idea of perfectly preserving metric properties, seeking instead to strike a balance between distortions, or to simply make things "look right". Most of these types of projections distort shape in the polar regions more than at the equator. These are some compromise projections:

- Robinson
- van der Grinten
- Miller cylindrical
- Winkel Tripel
- Buckminster Fuller's Dymaxion
- B. J. S. Cahill's Butterfly Map
- Kavrayskiy VII projection
- Wagner VI projection
- Chamberlin trimetric
- Oronce Finé's cordiform

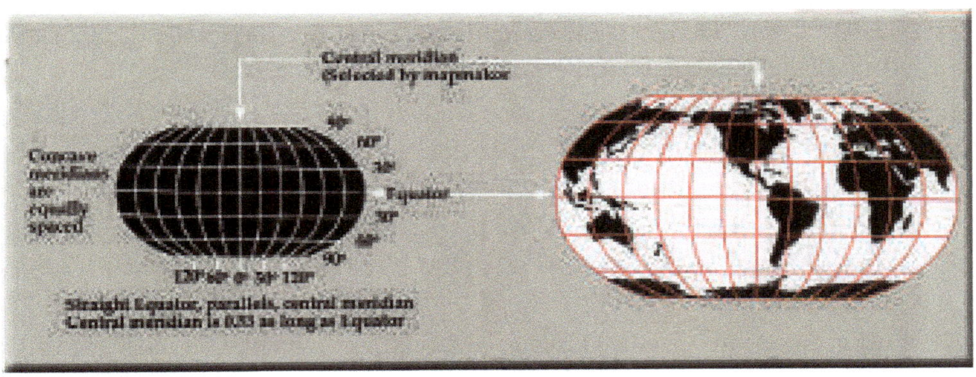

The Robinson projection was adopted by National Geographic magazine in 1988 but abandoned by them in about 1997 for the Winkel tripel.

GEOCODING

Geocoding is the computational process of transforming a postal address description to a location on the Earth's surface (spatial representation in numerical coordinates). Reverse geocoding, on the other hand, converts the inputted geographic coordinates to a description of a location, usually the name of a place or a postal address. Geocoding relies on a computer representation of the street network. Geocoding is sometimes used for conversion from ZIP codes or postal codes to coordinates, occasionally for the conversion of parcel identifiers to centroid coordinates.

Geocoding (verb): The act of transforming an address text into a valid spatial representation.

Geocoder (noun): A piece of software or a (web) service that implements a geocoding process i.e. a set of interrelated components in the form of operations, algorithms, and data sources that work together to produce a spatial representation for descriptive locational references.

Geocode (noun): A spatial representation of a descriptive locational reference.

The geographic coordinates representing locations often vary greatly in positional accuracy. Examples include building centroids, land parcels, street addresses, postal code centroids (e.g. ZIP codes, CEDEX), and Administrative Boundary Centroids.

History

Geocoding — a subset of Geographic Information System (GIS) spatial analysis — has been a subject of interest since the early 1960s.

1960s

In 1960, the first operational GIS — named the Canada Geographic Information System (CGIS) — was invented by Dr. Roger Tomlinson, who has since been acknowledged as the father of GIS. The CGIS was used to store and analyze data collected for the Canada Land Inventory, which mapped information about agriculture, wildlife, and forestry at a scale of 1:50,000, in order to regulate land capability for rural Canada. However, the CGIS lasted until the 1990s and was never available commercially.

On July 1, 1963, five-digit ZIP codes were introduced nationwide by the United States Post Office Department (USPOD). In 1983, nine-digit ZIP+4 codes were brought about as an extra identifier in more accurately locating addresses.

In 1964, the Harvard Laboratory for Computer Graphics and Spatial Analysis developed groundbreaking software code — e.g. GRID, and SYMAP — all of which were sources for commercial development of GIS.

In 1967, a team at the Census Bureau — including the mathematician James Corbett and Donald Cooke — invented Dual Independent Map Encoding (DIME) — the first modern vector mapping model — which ciphered address ranges into street network files and incorporated the "percent along" geocoding algorithm. Still in use by platforms such as Google Maps and MapQuest, the "percent along" algorithm denotes where a matched address is located along a reference feature as a percentage of the reference feature's total length. DIME was intended for the use of the United States Census Bureau, and it involved accurately mapping block faces, digitizing nodes representing street intersections, and forming spatial relationships. New Haven, Connecticut was the first city on Earth with a geocodable streets network database.

1980s

In the late 1970s, two main public domain geocoding platforms were in development: GRASS GIS and MOSS. The early 1980s saw the rise of many more commercial vendors of geocoding software, namely Intergraph, ESRI, CARIS, ERDAS, and MapInfo Corporation. These platforms merged the 1960s approach of separating spatial information with the approach of organizing this spatial information into database structures.

In 1986, Mapping Display and Analysis System (MIDAS) became the first desktop geocoding software, designed for the DOS operating system. Geocoding was elevated from the research department into the business world with the acquisition of MIDAS by MapInfo. MapInfo has since been acquired by Pitney Bowes, and has pioneered in merging geocoding with business intelligence; allowing location intelligence to provide solutions for the public and private sectors.

1990s

The end of the 20th century had seen geocoding become more user-oriented, especially via open-source GIS software. Mapping applications and geospatial data had become more accessible over the Internet.

Because the mail-out/mail-back technique was so successful in the 1980 Census, the U.S. Bureau of Census was able to put together a large geospatial database, using interpolated street geocoding. This database — along with the Census' nationwide coverage of households — allowed for the birth of TIGER (Topologically Integrated Geographic Encoding and Referencing).

Containing address ranges instead of individual addresses, TIGER has since been implemented in nearly all geocoding software platforms used today. By the end of the 1990 Census, TIGER "contained a latitude/longitude-coordinate for more than 30 million feature intersections and endpoints and nearly 145 million feature 'shape' points that defined the more than 42 million feature segments that outlined more than 12 million polygons."

TIGER was the breakthrough for "big data" geospatial solutions.

2000s

The early 2000s saw the rise of Coding Accuracy Support System (CASS) address standardization. The CASS certification is offered to all software vendors and advertising mailers who want the United States Postal Services (USPS) to assess the quality of their address-standardizing software. The annually renewed CASS certification is based on delivery point codes, ZIP codes, and ZIP+4 codes. Adoption of a CASS certified software by software vendors allows them to receive discounts in bulk mailing and shipping costs.

They can benefit from increased accuracy and efficiency in those bulk mailings, after having a certified database. In the early 2000s, geocoding platforms were also able to support multiple datasets.

In 2003, geocoding platforms were capable of merging postal codes with street data, updated monthly. This process became known as "conflation".

Beginning in 2005, geocoding platforms included parcel-centroid geocoding. Parcel-centroid geocoding allowed for a lot of precision in geocoding an address. For example, parcel-centroid allowed a geocoder to determine the centroid of a specific building or lot of land. Platforms were now also able to determine the elevation of specific parcels.

2005 also saw the introduction of the Assessor's Parcel Number (APN). A jurisdiction's tax assessor was able to assign this number to parcels of real estate. This allowed for proper identification and record-keeping. An APN is important for geocoding an area which is covered by a gas or oil lease, and indexing property tax information provided to the public.

In 2006, Reverse Geocoding and reverse APN lookup were introduced to geocoding platforms. This involved geocoding a numerical point location — with a longitude and latitude — to a textual, readable address.

2008 and 2009 saw the growth of interactive, user-oriented geocoding platforms — namely MapQuest, Google Maps, Bing Maps, and Global Positioning Systems (GPS). These platforms were made even more accessible to the public with the simultaneous growth of the mobile industry, specifically smartphones.

2010s

This current decade has seen vendors fully supporting geocoding and reverse geocoding globally. Cloud-based geocoding application programming interface (API) and on-premise geocoding has allowed for a greater match rate, greater precision, and greater speed. There is now a popularity in the idea of geocoding being able to influence business decisions. This is the integration between the geocoding process and business intelligence.

The future of geocoding also involves three-dimensional geocoding, indoor geocoding, and multiple language returns for the geocoding platforms.

Geocoding Process

Geocoding is a task which involves multiple datasets and processes, all of which work together. A geocoder is made of two important components: a reference dataset and the geocoding algorithm. Each of these components are made up of sub-operations and sub-components. Without understanding how these geocoding processes work, it is difficult to make informed business decisions based on geocoding.

Input Data

Input data are the descriptive, textual information (address or building name) which the user wants to turn into numerical, spatial data (latitude and longitude) — through the process of geocoding.

Classification of Input Data

Input data is classified into two categories: relative input data and absolute input data.

Relative Input Data

Relative input data are the textual descriptions of a location which, alone, cannot output a spatial representation of that location. Such data outputs a relative geocode, which is dependent and geographically relative of other reference locations. An example of a relative geocode is address-interpolation using areal units or line vectors. "Across the street from the Empire State Building" is an example of a relative input data. The location being sought cannot be determined without identifying the Empire State Building. Geocoding platforms often do not support such relative locations, but advances are being made in this direction.

Absolute Input Data

Absolute input data are the textual descriptions of a location which, alone, can output a spatial representation of that location. This data type outputs an absolute known location independently of other locations. For example, USPS ZIP codes; USPS ZIP+4 codes; complete and partial postal addresses; USPS PO boxes; rural routes; cities; counties; intersections; and named places can all be referenced in a data source absolutely.

When there is a lot of variability in the way addresses can be represented — such as too much input data or too little input data — geocoders use address normalization and address standardization in order to resolve this problem.

Processing of Input Data

Address Interpolation

A simple method of geocoding is address interpolation. This method makes use of data from a street geographic information system where the street network is already mapped within the geographic coordinate space. Each street segment is attributed with address ranges (e.g. house numbers from one segment to the next). Geocoding takes an address, matches it to a street and specific segment (such as a block, in towns that use the "block" convention). Geocoding then interpolates the position of the address, within the range along the segment.

Example

Take for example: 742 Evergreen Terrace

Let's say that this segment (for instance, a block) of Evergreen Terrace runs from 700 to 799. Even-numbered addresses fall on the east side of Evergreen Terrace, with odd-numbered addresses on the west side of the street. 742 Evergreen Terrace would (probably) be located slightly less than halfway up the block, on the east side of the street. A point would be mapped at that location along the street, perhaps offset a distance to the east of the street centerline.

Complicating Factors

However, this process is not always as straightforward as in this example. Difficulties arise when

- distinguishing between ambiguous addresses such as 742 Evergreen Terrace and 742 W Evergreen Terrace.
- attempting to geocode new addresses for a street that is not yet added to the geographic information system database.

While there might be 742 Evergreen Terrace in Springfield, there might also be a 742 Evergreen Terrace in Shelbyville. Asking for the city name (and state, province, country, etc. as needed) can solve this problem. Boston, Massachusetts has multiple "100 Washington Street" locations because several cities have been annexed without changing street names, thus requiring use of unique postal codes or district names for disambiguation. Geocoding accuracy can be greatly improved by first utilizing good address verification practices. Address verification will confirm the existence of the address and will eliminate ambiguities. Once the valid address is determined, it is very easy to geocode and determine the latitude/longitude coordinates. Finally, several caveats on using interpolation:

- The typical attribution of a street segment assumes that all even numbered parcels are on one side of the segment, and all odd numbered parcels are on the other. This is often not true in real life.
- Interpolation assumes that the given parcels are evenly distributed along the length of the segment. This is almost never true in real life; it is not uncommon for a geocoded address to be off by several thousand feet.
- Interpolation also assumes that the street is straight. If a street is curved then the geocoded location will not necessarily fit the physical location of the address.
- Segment Information (esp. from sources such as TIGER) includes a maximum upper bound for addresses and is interpolated as though the full address range is used. For example, a segment (block) might have a listed range of 100-199, but the last address at the end of the block is 110. In this case, address 110 would be geocoded to 10% of the distance down the segment rather than near the end.
- Most interpolation implementations will produce a point as their resulting address location. In reality, the physical address is distributed along the length of the segment, i.e. consider geocoding the address of a shopping mall - the physical

lot may run a distance along the street segment (or could be thought of as a two-dimensional space-filling polygon which may front on several different streets — or worse, for cities with multi-level streets, a three-dimensional shape that meets different streets at several different levels) but the interpolation treats it as a singularity.

A very common error is to believe the accuracy ratings of a given map's geocodable attributes. Such accuracy currently touted by most vendors has no bearing on an address being attributed to the correct segment, being attributed to the correct side of the segment, nor resulting in an accurate position along that correct segment. With the geocoding process used for U.S. Census TIGER datasets, 5-7.5% of the addresses may be allocated to a different census tract, while a study of Australia's TIGER-like system found that 50% of the geocoded points were mapped to the wrong property parcel. The accuracy of geocoded data can also have a bearing on the quality of research that can be done using this data. One study by a group of Iowa researchers found that the common method of geocoding using TIGER datasets as described above, can cause a loss of as much as 40% of the power of a statistical analysis. An alternative is to use orthophoto or image coded data such as the Address Point data from Ordnance Survey in the UK, but such datasets are generally expensive. Because of this, it is quite important to avoid using interpolated results except for non-critical applications, such as pizza delivery. Interpolated geocoding is usually not appropriate for making authoritative decisions, for example if life safety will be affected by that decision. Emergency services, for example, do not make an authoritative decision based on their interpolations; an ambulance or fire truck will always be dispatched regardless of what the map says.

Other Techniques

In rural areas or other places lacking high quality street network data and addressing, GPS is useful for mapping a location. For traffic accidents, geocoding to a street intersection or midpoint along a street centerline is a suitable technique. Most highways in developed countries have mile markers to aid in emergency response, maintenance, and navigation. It is also possible to use a combination of these geocoding techniques — using a particular technique for certain cases and situations and other techniques for other cases. In contrast to geocoding of structured postal address records, toponym resolution maps place names in unstructured document collections to their corresponding spatial footprints.

Research

Research has introduced a new approach to the control and knowledge aspects of geocoding, by using an agent-based paradigm. In addition to the new paradigm for geocoding, additional correction techniques and control algorithms have been developed. The approach represents the geographic elements commonly found in addresses as individual agents. This provides a commonality and duality to control and geographic representation. In addition to scientific publication, the new approach and subsequent prototype gained national media coverage in Australia. The research was conducted at Curtin University in Perth, Western Australia.

Uses

Geocoded locations are useful in many GIS analysis, cartography, decision making workflow, transaction mash-up, or injected into larger business processes. On the web, geocoding is used in services like routing and local search. Geocoding, along with GPS provides location data for geotagging media, such as photographs or RSS items.

Privacy Concerns

The proliferation and ease of access to geocoding (and reverse-geocoding) services raises privacy concerns. For example, in mapping crime incidents, law enforcement agencies aim to balance the privacy rights of victims and offenders, with the public's right to know. Law enforcement agencies have experimented with alternative geocoding techniques that allow them to mask a portion of the locational detail (e.g., address specifics that would lead to identifying a victim or offender). As well, in providing online crime mapping to the public, they also place disclaimers regarding the locational accuracy of points on the map, acknowledging these location masking techniques, and impose terms of use for the information.

WEB MAPPING

Web mapping is the process of using maps delivered by geographical information systems (GIS). A web map on the World Wide Web is both served and consumed, thus web mapping is more than just web cartography, it is a service by which consumers may choose what the map will show. Web GIS emphasizes geodata processing aspects more involved with design aspects such as data acquisition and server software architecture such as data storage and algorithms, than it does the end-user reports themselves.

The terms web GIS and web mapping remain somewhat synonymous. Web GIS uses web maps, and end users who are web mapping are gaining analytical capabilities. The term location-based services refers to web mapping consumer goods and services. Web mapping usually involves a web browser or other user agent capable of client-server interactions. Questions of quality, usability, social benefits, and legal constraints are driving its evolution.

The advent of web mapping can be regarded as a major new trend in cartography. Until recently cartography was restricted to a few companies, institutes and mapping agencies, requiring relatively expensive and complex hardware and software as well as skilled cartographers and geomatics engineers.

Web mapping has brought many geographical datasets, including free ones generated by OpenStreetMap and proprietary datasets owned by Navteq, Google, Waze, and others. A range of free software to generate maps has also been conceived and implemented alongside proprietary tools like ArcGIS. As a result, the barrier to entry for serving maps on the web has been lowered.

Types of Web Maps

A first classification of web maps has been made by Kraak in 2001. He distinguished static and dynamic web maps and further distinguished interactive and view only web maps. Today there an increased number of dynamic web maps types, and static web map sources.

Analytical Web Maps

Analytical web maps offer GIS analysis. The geodata can be a static provision, or needs updates. The borderline between analytical web maps and web GIS is fuzzy. Parts of the analysis can be carried out by the GIS geodata server. As web clients gain capabilities processing is distributed.

Animated and Realtime

Realtime maps show the situation of a phenomenon in close to realtime (only a few seconds or minutes delay). They are usually animated. Data is collected by sensors and the maps are generated or updated at regular intervals or on demand.

Animated maps show changes in the map over time by animating one of the graphical or temporal variables. Technologies enabling client-side display of animated web maps include scalable vector graphics (SVG), Adobe Flash, Java, QuickTime, and others. Web maps with real-time animation include weather maps, traffic congestion maps and vehicle monitoring systems.

CartoDB launched an open source library, Torque, which enables the creation of dynamic animated maps with millions of records. Twitter uses this technology to create maps to reflect how users reacted to news and events worldwide.

Collaborative Web Maps

Collaborative maps are a developing potential. In proprietary or open source collaborative software, users collaborate to create and improve the web mapping experience. Some collaborative web mapping projects are:

- Google Map Maker
- Here Map Creator
- OpenStreetMap
- WikiMapia
- meta:Maps - a survey of Wikimedia web mapping proposals

Online Atlases

The traditional atlas goes through a remarkably large transition when hosted on the web. Atlases can cease their printed editions or offer printing on demand. Some atlases also offer raw data downloads of the underlying geospatial data sources.

Static Web Maps

Static web pages are view only without animation or interactivity. These files are created once, often manually, and infrequently updated. Typical graphics formats for static web maps are PNG, JPEG, GIF, or TIFF (e.g., drg) for raster files, SVG, PDF or SWF for vector files. These include scanned paper maps not designed as screen maps. Paper maps have a much higher resolution and information density than typical computer displays of the same physical size, and might be unreadable when displayed on screens at the wrong resolution.

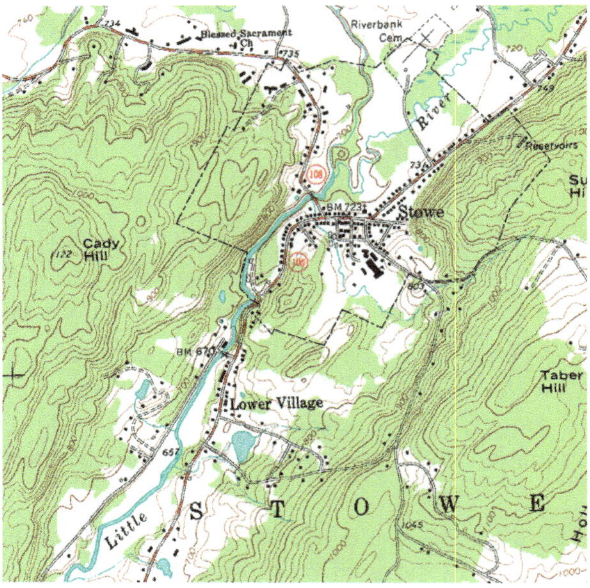

A USGS DRG - a static map

Evolving Paper Cartography

Compared to traditional techniques, mapping software has many advantages. The disadvantages are also stated.

- Web maps can easily deliver up to date information. If maps are generated automatically from databases, they can display information in almost realtime. They don't need to be printed, mastered and distributed. Examples:
    - A map displaying election results, as soon as the election results become available.
    - A traffic congestion map using traffic data collected by sensor networks.
    - A map showing the current locations of mass transit vehicles such as buses or trains, allowing patrons to minimize their waiting time at stops or stations, or be aware of delays in service.
    - Weather maps, such as NEXRAD.

- Software and hardware infrastructure for web maps is cheap. Web server hardware is cheaply available and many open source tools exist for producing web maps. Geodata, on the other hand, is not; satellites and fleets of automobiles use expensive equipment to collect the information on an ongoing basis. Perhaps owing to this, many people are still reluctant to publish geodata, especially in places where geodata are expensive. They fear copyright infringements by other people using their data without proper requests for permission.

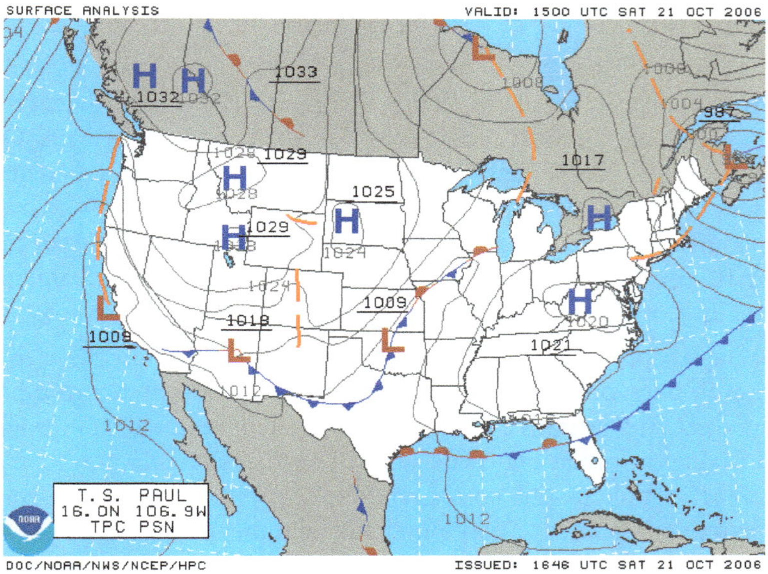

A surface weather analysis for the United States on October 21, 2006.

- Product updates can easily be distributed. Because web maps distribute both logic and data with each request or loading, product updates can happen every time the web user reloads the application. In traditional cartography, when dealing with printed maps or interactive maps distributed on offline media (CD, DVD, etc.), a map update takes serious efforts, triggering a reprint or remastering as well as a redistribution of the media. With web maps, data and product updates are easier, cheaper, and faster, and occur more often. Perhaps owing to this, many web maps are of poor quality, both in symbolization, content and data accuracy.
- Web maps can combine distributed data sources. Using open standards and documented APIs one can integrate (mash up) different data sources, if the projection system, map scale and data quality match. The use of centralized data sources removes the burden for individual organizations to maintain copies of the same data sets. The downside is that one has to rely on and trust the external data sources. In addition, with detailed information available and the combination of distributed data sources, it is possible to find out and combine a lot of private and personal information of individual persons. Properties and estates of individuals are now accessible through high resolution aerial and satellite images throughout the world to anyone.

- Web maps allow for personalization. By using user profiles, personal filters and personal styling and symbolization, users can configure and design their own maps, if the web mapping systems supports personalization. Accessibility issues can be treated in the same way. If users can store their favourite colors and patterns they can avoid color combinations they can't easily distinguish (e.g. due to color blindness). Despite this, as with paper, web maps have the problem of limited screen space, but more so. This is in particular a problem for mobile web maps; the equipment carried usually has a very small screen, making it less likely that there is room for personalisation.
- Web maps enable collaborative mapping similar to web mapping technologies such as DHTML/Ajax, SVG, Java, Adobe Flash, etc. enable distributed data acquisition and collaborative efforts. Examples for such projects are the OpenStreetMap project or the Google Earth community. As with other open projects, quality assurance is very important, however, and the reliability of the internet and web server infrastructure is not yet good enough. Especially if a web map relies on external, distributed data sources, the original author often cannot guarantee the availability of the information.
- Web maps support hyperlinking to other information on the web. Just like any other web page or a wiki, web maps can act like an index to other information on the web. Any sensitive area in a map, a label text, etc. can provide hyperlinks to additional information. As an example a map showing public transport options can directly link to the corresponding section in the online train time table. However, development of web maps is complicated enough as it is: Despite the increasing availability of free and commercial tools to create web mapping and web GIS applications, it is still a more complex task to create interactive web maps than to typeset and print images. Many technologies, modules, services and data sources have to be mastered and integrated The development and debugging environments of a conglomerate of different web technologies is still awkward and uncomfortable.

History of Web Mapping

Event types

- ***Cartography-related events***
- **Technical events directly related to web mapping**
- General technical events
- Events relating to Web standards

This section contains some of the milestones of web mapping, online mapping services and atlases.

- 1989: Birth of the WWW, WWW invented at CERN for the exchange of research documents.

- 1993: Xerox PARC Map Viewer, The first mapserver based on CGI/Perl, allowed reprojection styling and definition of map extent.
- 1994: The World Wide Earthquake Locator, the first interactive web mapping mashup was released, based on the Xerox PARC map view.
- 1994: The National Atlas of Canada, The first version of the National Atlas of Canada was released. Can be regarded as the first online atlas.
- 1995: The Gazetteer for Scotland, The prototype version of the Gazetteer for Scotland was released. The first geographical database with interactive mapping.
- 1995: MapGuide, First introduced as Argus MapGuide.
- 1996: Center for Advanced Spatial Technologies Interactive Mapper, Based on CGI/C shell/GRASS would allow the user to select a geographic extent, a raster base layer, and number of vector layers to create personalized map.
- 1996: Mapquest, The first popular online Address Matching and Routing Service with mapping output.
- 1996: MultiMap, The UK-based MultiMap website launched offering online mapping, routing and location based services. Grew into one of the most popular UK web sites.
- 1996: Geomedia WebMap 1.0, First version of Geomedia WebMap, already supports vector graphics through the use of ActiveCGM.
- 1996: MapGuide, Autodesk acquired Argus Technologies.and introduced Autodesk MapGuide 2.0.

National Atlas of the United States logo

- 1997: US Online National Atlas Initiative, The USGS received the mandate to coordinate and create the online National Atlas of the United States of America .
- 1997: UMN MapServer 1.0, Developed at the University of Minnesota (UMN) as Part of the NASA ForNet Project. Grew out of the need to deliver remote sensing data across the web for foresters.
- 1997: GeoInfoMapper - GeoInfo Solutions developed the first Java GIS Applet called 'JavaMap'. The application supported the export and conversion of MapInfo data for display in the thematic mapping tool for the web. GeoinfoMapper was demonstrated at the Victoria Computer Show in 1997 and referenced in the Universal Locator project at UC Berkeley School of Information.
- 1998: Terraserver USA, A Web Map Service serving aerial images (mainly b+w) and USGS DRGs was released. One of the first popular WMS. This service is a joint effort of USGS, Microsoft and HP.
- 1998: UMN MapServer 2.0, Added reprojection support (PROJ.4).
- 1998: MapObjects Internet Map Server, ESRI's entry into the web mapping business.
- 1999: National Atlas of Canada, 6th edition, This new version was launched at the ICA 1999 conference in Ottawa. Introduced many new features and topics.

Is being improved gradually, since then, and kept up-to-date with technical advancements.

- 2000: ArcIMS 3.0, The first public release of ESRI's ArcIMS.
- 2000: ESRI Geography Network, ESRI founded Geography Network to distribute data and web map services.
- 2000: UMN MapServer 3.0, Developed as part of the NASA TerraSIP Project. This is also the first public, open source release of UMN Mapserver. Added raster support and support for TrueType fonts (FreeType).
- 2001: GeoServer, starts of the GeoServer project (Geoserver History)
- 2001: MapScript 1.0 for UMN MapServer, Adds a lot of flexibility to UMN MapServer solutions.
- 2001: Tirolatlas, A highly interactive online atlas, the first to be based on the SVG standard.
- 2002: UMN MapServer 3.5, Added support for PostGIS and ArcSDE. Version 3.6 adds initial OGC WMS support.
- 2002: ArcIMS 4.0, Version 4 of the ArcIMS web map server.

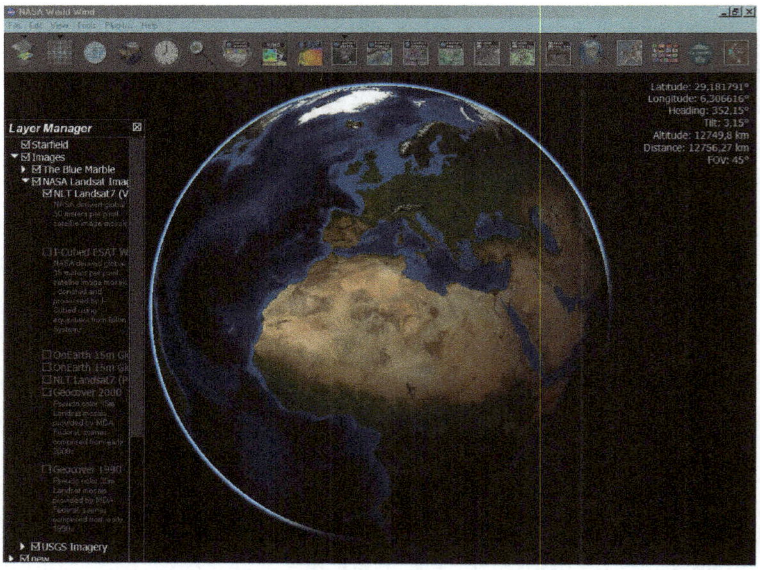

Screenshot from NASA World Wind

- 2003: NASA World Wind, NASA World Wind Released. An open virtual globe that loads data from distributed resources across the internet. Terrain and buildings can be viewed 3 dimensionally. The (XML based) markup language allows users to integrate their own personal content. This virtual globe needs special software and doesn't run in a web browser.
- 2003: UMN MapServer 4.0, Adds 24bit raster output support and support for PDF and SWF.

- 2004: OpenStreetMap, an open source, open content world map founded by Steve Coast.
- 2005: Google Maps, The first version of Google Maps. Based on raster tiles organized in a quad tree scheme, data loading done with XMLHttpRequests. This mapping application became highly popular on the web, also because it allowed other people to integrate google map services into their own website.
- 2005: UMN MapServer introduced as open source by the Open Source Geospatial Foundation (OSGeo). UMN MapServer 4.6, Adds support for SVG.
- 2005: MapGuide Open Source introduced as open source by Autodesk
- 2005: Google Earth, The first version of Google Earth was released building on the virtual globe metaphor. Terrain and buildings can be viewed 3 dimensionally. The KML (XML based) markup language allows users to integrate their own personal content. This virtual globe needs special software and doesn't run in a web browser.
- 2005: OpenLayers, the first version of the open source Javascript library OpenLayers.
- 2006: WikiMapia Launched
- 2009: Nokia makes Ovi Maps free on its smartphones.
- 2010: MapBox is founded
- 2012: Apple removes Google Maps as the default mapping app and replaces it with its own mapping app
- 2013: MapBox announces Vector Tiles for MapBox Streets

Web Mapping Technologies

Web mapping technologies require both server and client side applications. The following is a list of technologies utilized in web mapping.

- Spatial databases are usually object relational databases enhanced with geographic data types, methods and properties. They are necessary whenever a web mapping application has to deal with dynamic data (that changes frequently) or with huge amount of geographic data. Spatial databases allow spatial queries, sub selects, reprojections, and geometry manipulations and offer various import and export formats. PostGIS is a prominent example; it is open source. MySQL also implements some spatial features. Oracle Spatial, Microsoft SQL Server (with the spatial extensions), and IBM DB2 are the commercial alternatives. The Open Geospacial Consortium's (OGC) specification "Simple Features" is a standard geometry data model and operator set for spatial databases. Part 2 of the specification defines an implementation using SQL.
- Tiled web maps display rendered maps made up of raster image "tiles".
- Vector tiles are also becoming more popular-- Google and Apple have both transitioned to vector tiles. Mapbox.com also offers vector tiles. This new

style of web mapping is resolution independent, and also has the advantage of dynamically showing and hiding features depending on the interaction.

- WMS servers generate maps using parameters for user options such as the order of the layers, the styling and symbolization, the extent of the data, the data format, the projection, etc. The OGC standardized these options. Another WMS server standard is the Tile Map Service. Standard image formats include PNG, JPEG, GIF and SVG. Open source WMS Servers include UMN Mapserver, GeoServer and Mapnik. Commercial alternatives exist from most commercial GIS vendors, such as ESRI ArcIMS and CadCorp.

GEOGRAPHIC INFORMATION SYSTEMS IN GEOSPATIAL INTELLIGENCE

Geographic information systems (GIS) play a constantly evolving role in geospatial intelligence (GEOINT) and United States national security. These technologies allow a user to efficiently manage, analyze, and produce geospatial data, to combine GEOINT with other forms of intelligence collection, and to perform highly developed analysis and visual production of geospatial data. Therefore, GIS produces up-to-date and more reliable GEOINT to reduce uncertainty for a decision maker. Since GIS programs are Web-enabled, a user can constantly work with a decision maker to solve their GEOINT and national security related problems from anywhere in the world. There are many types of GIS software used in GEOINT and national security, such as Google Earth, ERDAS IMAGINE, GeoNetwork opensource, and Esri ArcGIS.

Background

A user can enter different kinds of data in map form into a GIS to begin their analysis, such as United States Geological Survey (USGS) digital line graph data, contour lines, elevation maps, topographic maps, geologic maps, and satellite imagery. A user can also convert digital information into forms that a GIS can identify and utilize, such as census tabular data or Microsoft Excel files. Users can easily capture digital data in a GIS. If the data is not digital, then users will need to employ various techniques to capture the data, such as digitizing maps by hand-tracing with a computer mouse, utilizing a digitizing tablet to collect feature coordinates, using electronic scanners, or uploading Global Positioning System (GPS) coordinates.

GIS applies to the geographical facets of various aspects of everyday life, such as transportation, logistics, medicine, marketing, sociology, ecology, pure and applied sciences, emergency management, and criminology. GIS is also utilized in all three areas of intelligence: national security intelligence, law enforcement intelligence, and competitive intelligence

Geospatial Intelligence (GEOINT)

GEOINT, known previously as imagery intelligence (IMINT), is an intelligence collection discipline that applies to national security intelligence, law enforcement intelligence, and competitive intelligence. For example, an analyst can use GEOINT to identify the route of least resistance for a military force in a hostile country, to discover a pattern in the locations of reported burglaries in a neighborhood, or to generate a map and comparison of failing businesses that a company is likely to purchase. GEOINT is also the geospatial product of a process that is focused externally, designed to reduce the level of uncertainty for a decisionmaker, and that uses information derived from all sources. The National Geospatial-Intelligence Agency (NGA), who has overall responsibility for GEOINT in the U.S. Intelligence Community (IC), defines GEOINT as "information about any object—natural or man-made—that can be observed or referenced to the Earth, and has national security implications."

Some of the sources of collected imagery information for GEOINT are imagery satellites, cameras on airplanes, Unmanned Aerial Vehicles (UAV) and drones, handheld cameras, maps, or GPS coordinates. Recently the NGA and IC have increased the use of commercial satellite imagery for intelligence support, such as the use of the IKONOS, Landsat, or SPOT satellites. These sources produce digital imagery via electro-optical systems, radar, infrared, visible light, multispectral, or hyperspectral imageries.

The advantages of GEOINT are that imagery is easily consumable and understood by a decisionmaker, has low human life risk, displays the capabilities of a target and its geographical relationship to other objects, and that analysts can use imagery world-wide in a short time. On the other hand, the disadvantages of GEOINT are that imagery is only a snapshot of a moment in time, can be too compelling and lead to ill-informed decisions that ignore other intelligence, is static and vulnerable to deception and decoys, does not depict the intentions of a target, and is expensive and subject to environmental problems.

GIS use in GEOINT and National Security Intelligence

A majority of national security intelligence decisions involve geography and GEOINT. GIS allows the user to capture, manage, exploit, analyze, and visualize geographically referenced information, physical features, and other geospatial data. GIS is thus a critical infrastructure for the GEOINT and national security community in manipulating and interpreting spatial knowledge in an information system. GIS extracts real world geographic or other information into datasets, maps, metadata, data models, and workflow models within a geodatabase that is used to solve GEOINT-related problems. GIS provides a structure for map and data production that allows a user to add other data sources, such as satellite or UAV imagery, as new layers to a geodatabase. The geodatabase can be disseminated and operated across any network of associated users (i.e. from the GEOINT analyst to the warfighter) and engenders a common spatial capability for all defense and intelligence domains.

The map and chart production agency and imagery intelligence agency, the principal two agencies of GEOINT, use GIS to efficiently work together to solve decisionmaker's geospatial questions, to communicate effectively between their unique departments, and to provide constantly updated, accurate GEOINT to their national security and warfighter domains.

Another important aspect of GIS is its ability to fuse geospatial data with other forms of intelligence collection, such as signals intelligence (SIGINT), measurement and signature intelligence (MASINT), human intelligence (HUMINT), or open source intelligence (OSINT). A GIS user can incorporate and fuse all of these types of intelligence into applications that provide corroborated GEOINT throughout an organization's information system.

GIS enables efficient management of geospatial data, the fusion of geospatial data with other forms of intelligence collection, and advanced analysis and visual production of geospatial data. This produces faster, corroborated, and more reliable GEOINT that aims to reduce uncertainty for a decisionmaker.

Roles

- Data and map production
- Data fusion, data discovery through metadata catalogs, and data dissemination through Web portals and browsers
- Analysis and exploitation of collected imagery or intelligence
- SIGINT, GEOINT, MASINT, and other sensor analysis
- Fusion of multiple forms of intelligence collection
- Collaborative planning and efficient workflow management between decisionmakers, analysts, consumers, and warfighters
- Suitability and temporal analysis
- Stewardship: Geospatial intelligence

Related Esri Products

Distributed Geospatial Intelligence Network (DGInet)

The DGInet technology allows military and national security intelligence customers to access large multi-terabyte databases through a common Web-based interface. This gives the users the capability to quickly and easily identify, overlay, and fuse georeferenced data from various sources to create maps or support geospatial analysis. Esri designed the technology for inexperienced GIS users of national security intelligence and defense organizations in order to provide a Web-based enterprise solution for publishing, distributing, and exploiting GEOINT data among designated organizations. According to Esri, the DGInet technology "uses thin clients to search massive amounts of geospatial and intelligence data using low-bandwidth Web services for data discovery, dissemination, and horizontal fusion of data and products."

PLTS for ArcGIS Specialized Solutions

PLTS for ArcGIS Specialized Solutions is a group of software applications that extends ArcGIS to facilitate database driven cartographic production for geospatial and mapping agencies, nautical and aeronautical chart production, foundation mapping, and defense mapping requirements. The collection of software applications includes Esri Production Mapping, Esri Nautical Solution, Esri Aeronautical Solution, and Esri Defense Mapping programs that provide quality control, easier and consistent map production, database sharing, and efficient workflow management for each program's specific type of mapping or charting.

Geoprocessing

Geoprocessing is based on a framework of data transformation in GIS and is a collection of hundreds of GIS tools that manipulate geospatial or other data in GIS. A geoprocessing tool performs an operation (often the name of the tool, such as "Clip") on an existing GIS dataset and produces a new dataset as a result of the utilized tool. GIS users utilize these tools to create a workflow model that quickly and easily transforms raw data into the desired product.

In GEOINT, users employ geoprocessing in similar ways. They can make geoprocessing tools resemble analytic techniques to transform large amounts of data into actionable information. In national security intelligence and defense organizations, geoprocessing notifies users to events occurring in specific areas of interest and enables domain-specific analysis applications, such as radio frequency analysis, terrain analysis, and network analysis.

Tracking Analyst and Tracking Server

The ArcGIS Tracking Analyst extension enables the user to create time series visualizations to analyze time and location sensitive information. It creates a visible path from incorporated data that shows movement through space and time. The program allows the national security intelligence or defense user to track assets (such as vehicles or personnel), monitor sensors, visualize change over time, play back events, and analyze historical or real-time temporal data.

The Tracking Server program is an Esri enterprise technology that integrates real-time data with GIS to disseminate information quickly and easily to decisionmakers. This program enables the user to obtain data in any format and transmit it to the necessary consumer or ArcGIS Tracking Analyst user, to conduct filters or alerts on specific attributes of incoming data or global positions, and to log data into ArcGIS Server for efficient project management and information sharing.

When Tracking Server and ArcGIS Tracking Analyst are used together, a user can monitor changes in data as they occur in real-time. A national security intelligence or defense user can subscribe to real-time data over the Internet from GPS and custom data feeds to support GEOINT requirements, such as fleet management or target tracking.

ArcGIS Military Analyst

The ArcGIS Military Analyst extension incorporates display and analysis tools that allow the use and production of vector and raster products, line-of-sight analysis, hillshade analysis, terrain analysis, and Military Grid Reference System (MGRS) conversion. This program also provides a basis for command, control, and intelligence (C2I) systems. National security intelligence and defense organizations use ArcGIS Military Analyst extension to integrate geospatial data with other defense data, analyze digital terrains, and prepare for battle. This program also enables such users to manage and analyze geospatial data and relationships between mission planning, logistics, and C2I.

Military Overlay Editor (MOLE)

MOLE is a set of command components that enables national security intelligence and defense users to easily create, display, and edit U.S. Department of Defense MIL-STD-2525B and the North Atlantic Treaty OrganizationAPP-6A military symbology in a map. This allows for easier and faster identification, understanding, and movement of ally and hostile forces on a map by combining GIS spatial analysis techniques with common military symbols. MOLE provides a clearer visualization of mission planning and goals for the decisionmaker, and allows a user to import, locate, and display order of battle databases.

Grid Manager

Grid Manager enables the national security intelligence or defense user to create accurate, realistic grids that contain geographic location indicators based on specified shapes, scales, coordinate systems, and units. This program allows the user to create multiple grids, graticules, and borders for such map products as MGRS coordinates and tourist, topographic, parcel, street, nautical, and aeronautical maps.

GIS use in the National Geospatial-Intelligence Agency (NGA)

The NGA uses GIS products to create digital nautical, aeronautical, and topographic charts and maps, to perform geotechnical and coordinate system analysis, and to help solve a large variety of national security and military problems. Since the NGA is a U.S. Department of Defense combat support agency and a member of the IC, it uses GIS to produce precise, up-to-date GEOINT for members of the U.S. Armed Forces, the IC, and other government agencies. Web-enabled GIS applications allow for fast, efficient sharing and disseminating of geospatial data, products, and intelligence from the NGA to its allies, warfighters, partners, and other agencies across the World Wide Web. The NGA and Esri have successfully collaborated on providing timely, accurate, and relevant GEOINT in support of U.S. national security for the past 20 years.

The NGA has created a grouping of web-based capabilities called GEOINT Online. This program allows a user to search and access all NGA GEOINT documents from wherever they are stored and from wherever the user is. GEOINT Online provides quick, easy, and reliable access to current NGA intelligence products, changes in activities or regions, information from analyst's blogs and Intellipedia, geospatial imagery, maps and charts, major GIS commercial software packages, and GIS combinations of these products. A user can also edit and format existing NGA/GIS products and maps to create, print, and download new products that fulfill current decision maker requirements. Ultimately, this results in the faster production of timely and relevant GEOINT data. This program allowed the NGA to change its focus from simply generating cartographic products to providing updated, accurate GEOINT to support the national security and military requirements of its customers.

REFERENCES

1. "Geographic Information Systems as an Integrating Technology: Context, Concepts, and Definitions". Kenneth E. Foote and Margaret Lynch, The Geographer's Craft Project, Department of Geography, The University of Colorado at Boulder. Retrieved 21 Apr 2015.
2. Clarke, K. C., 1986. Advances in geographic information systems, computers, environment and urban systems, Vol. 10, pp. 175–184.
3. Maliene V, Grigonis V, Palevičius V, Griffiths S (2011). "Geographic information system: Old principles with new capabilities". Urban Design International. pp. 1–6. doi:10.1057/udi.2010.25.
4. Goodchild, Michael F (2010). "Twenty years of progress: GIScience in 2010". Journal of Spatial Information Science. doi:10.5311/JOSIS.2010.1.2.
5. "The 50th Anniversary of GIS". ESRI. Retrieved 18 April 2013.
6. "Roger Tomlinson". UCGIS. 2014-02-21. Retrieved 2015-12-16.
7. "Rapport sur la marche et les effets du choléra dans Paris et le département de la Seine. Année 1832". Gallica. Retrieved 10 May 2012.
8. Fitzgerald, Joseph H. "Map Printing Methods". Archived from the original on 2007-06-04. Retrieved 2007-06-09.
9. "GIS Hall of Fame – Roger Tomlinson". URISA. Retrieved 2007-06-09.
10. Lovison-Golob, Lucia. "Howard T. Fisher". Harvard University. Retrieved 2007-06-09.
11. "Open Source GIS History – OSGeo Wiki Editors". Retrieved 2009-03-21.
12. Fu, P., and J. Sun. 2010. Web GIS: Principles and Applications. ESRI Press. Redlands, CA. ISBN 1-58948-245-X.
13. Tim Foresman 1997 The History of GIS (Geographic Information Systems): Perspectives from the Pioneers. (Prentice Hall Series in Geographic Information Science) Prentice Hall PTR; 1st edition (November 10, 1997), 416 p.
14. The Remarkable History of GIS - Geographical Information Systems."The Remarkable History of GIS". Retrieved 2015-05-05.
15. Cowen 1988 "GIS VERSUS CAD VERSUS DBMS: WHAT ARE THE DIFFERENCES ?" PHOTOGRAMMETRIC ENGINEERING & REMOTE SENSING Vol. 54, No.11, November 1988, pp. 1551–1555. http://funk.on.br/esantos/doutorado/GEO/igce/DBMS.pdf last retrieved 9/17/2010.
16. http://www.fgdc.gov/standards/projects/FGDC-standards-projects/accuracy/part3/chapter3

17. "NJGIN's Information Warehouse". Njgin.state.nj.us. Retrieved 2012-05-13.
18. "Aeryon Announces Version 5 of the Aeryon Scout System | Aeryon Labs Inc". Aeryon.com. 2011-07-06. Retrieved 2012-05-13.
19. Winther, Rasmus G. (2014) "Mapping Kinds in GIS and Cartography" in Natural Kinds and Classification in Scientific Practice, edited by C. Kendig http://philpapers.org/archive/WINMKI.pdf
20. "Geospatial Analysis – a comprehensive guide. 2nd edition © 2006–2008 de Smith, Goodchild, Longley. Spatialanalysisonline.com. Retrieved 2012-05-13.
21. Chang, K. T. (2008). Introduction to Geographical Information Systems. New York: McGraw Hill. p. 184.
22. Longley, P. A.; Goodchild, M. F.; McGuire, D. J.; Rhind, D. W. (2005). Analysis of errors of derived slope and aspect related to DEM data properties. Geographic Information Systems and Science. West Sussex, England: John Wiley and Sons. p. 328.
23. Chang, K. T. (1989). A comparison of techniques for calculating gradient and aspect from a gridded digital elevation model. International Journal of Geographical Information Science. 3 (4): 323–334. doi:10.1080/02693798908941519.
24. Jones, K.H. (1998). A comparison of algorithms used to compute hill slope as a property of the DEM. Computers and Geosciences. 24 (4): 315–323. Bibcode:1998CG.....24..315J. doi:10.1016/S0098-3004(98)00032-6.
25. Zhou, Q.; Liu, X. (2003). Analysis of errors of derived slope and aspect related to DEM data properties. Computers and Geosciences. 30: 269–378.
26. Hunter G. J.; Goodchild M. F. (1997). Modeling the uncertainty of slope and aspect estimates derived from spatial databases (PDF). Geographical Analysis. 29 (1): 35–49. doi:10.1111/j.1538-4632.1997.tb00944.x.
27. K. Calvert, J. M. Pearce, W.E. Mabee, Toward renewable energy geo-information infrastructures: Applications of GIScience and remote sensing that can build institutional capacity Renewable and Sustainable Energy Reviews 18, pp. 416–429 (2013). open access
28. Heywood I, Cornelius S, Carver S (2006). An Introduction to Geographical Information Systems (3rd ed.). Essex, England: Prentice Hall.
29. Greene, R.; Devillers, R.; Luther, J.E.; Eddy, B.G. (2011). GIS-based multi-criteria analysis. Geography Compass. 5/6: 412–432.
30. Ma, Y.; Guo, Y.; Tian, X.; Ghanem, M. (2011). Distributed Clustering-Based Aggregation Algorithm for Spatial Correlated Sensor Networks. IEEE Sensors Journal. 11 (3): 641. doi:10.1109/JSEN.2010.2056916.
31. Integrating GIS with SAP--The Imperative - ArcNews Spring 2009 Issue. www.esri.com. Retrieved 2016-07-13.
32. Geodesy—Wolfram Language Documentation. reference.wolfram.com. Retrieved 2016-07-13.
33. Off the Map | From Architectural Record and Greensource | Originally published in the March 2012 issues of Architectural Record and Greensource | McGraw-Hill

Construction - Continuing Education Center. Continuingeducation.construction.com. 2011-03-11. Retrieved 2012-05-13.

34. Arctic Sea Ice Extent is Third Lowest on Record.
35. OGC Members | OGC(R). Opengeospatial.org. Retrieved 2012-05-13.
36. Fonseca, Frederico; Sheth, Amit (2002). The Geospatial Semantic Web (PDF). UCGIS White Paper.
37. Fonseca, Frederico; Egenhofer, Max (1999). Ontology-Driven Geographic Information Systems. Proc. ACM International Symposium on Geographic Information Systems: 14–19.
38. Perry, Matthew; Hakimpour, Farshad; Sheth, Amit (2006). Analyzing Theme, Space and Time: an Ontology-based Approach (PDF). Proc. ACM International Symposium on Geographic Information Systems: 147–154.
39. Archived May 21, 2007, at the Wayback Machine.
40. Semantic Web for Earth and Environmental Terminology.
41. W3C Geospatial Incubator Group.
42. International Conferences on Geospatial Semantics.
43. Archived May 18, 2007, at the Wayback Machine.
44. Haque, Akhlaque (2001-05-01). GIS, Public Service, and the Issue of Democratic Governance. Public Administration Review. 61 (3): 259–265. doi:10.1111/0033-3352.00028. ISSN 1540-6210.
45. Haque, Akhlaque (2003). Information technology, GIS and democraticvalues: Ethical implications for IT professionals in public service. Ethics and Information Technology. doi:10.1023/A:1024986003350.
46. Monmonier, Mark (2005). Lying with Maps. Statistical Science. JSTOR 20061176.
47. Monmonier, Mark (1991). How to Lie with Maps. Chicago, IL: University of Chicago Press. ISBN 0226534219.
48. Haque, Akhlaque (2015). Surveillance, Transparency and Democracy: Public Administration in the Information Age. Tuscaloosa, AL: University of Alabama Press. pp. 70–73. ISBN 978-0817318772.

Earth Observation Satellites

Chapter Contents

- Introduction
- Group on Earth Observations
- Global Earth Observation System of Systems
- Earth Observation Satellite

By observing the Earth from space, satellites provide essential information on ocean, ice, land environments and the atmosphere. Earth-observation satellites help us monitor and protect our environment, manage our resources, and ensure the safety and security. Satellite imagery and expertise is also used to support global humanitarian efforts and sustainable development.

INTRODUCTION

Earth observation is the gathering of information about planet Earth's physical, chemical and biological systems via remote sensing technologies supplemented by earth surveying techniques, encompassing the collection, analysis and presentation of data. Earth observation is used to monitor and assess the status of, and changes in, the natural environment and the built environment. In recent years, Earth observation has become technologically increasingly sophisticated. It has also become more important due to the dramatic impact that modern human civilization is having on the planet Earth, and the need to minimize negative impacts along with the opportunities Earth observation provides to improve social and economic well-being.

It must be noted here that the term “Earth observation” (EO) is used in two ways, leading to confusion. In Europe, in particular, it has often been used to refer to satellite-based remote sensing, but the term is also used to refer to any form of observations of the Earth system, including in situ and airborne observations, for example. The Group on Earth Observations, which has over 100 member countries and over 100 participating organizations, uses EO in this broader sense. To add to the confusion, in the US, for example, the term “remote sensing” is often used to refer to satellite-based remote sensing, but sometimes used more broadly for observations using any form of remote sensing technology, including airborne sensors and even ground-based sensors such as cameras. Perhaps the least ambiguous term to use for satellite-based sensors would be “satellite remote sensing”, or SRS, an acronym which is gradually starting to appear in the literature.

Earth observations can include:

- numerical measurements taken by a thermometer, wind gauge, ocean buoy, altimeter or seismometer
- photos and radar or sonar images taken from ground or ocean-based instruments
- photos and radar images taken from remote-sensing satellites
- decision-support tools based on processed information, such as maps and models

Just as Earth observations consist of a wide variety of possible elements, they can be applied to a wide variety of possible uses. Some of the specific applications of Earth observations include:

- forecasting weather
- tracking biodiversity and wildlife trends
- measuring land-use change (such as deforestation)
- monitoring and responding to natural disasters, including fires, floods, earthquakes and tsunamis
- managing natural resources, such as energy, freshwater and agriculture
- addressing emerging diseases and other health risks
- predicting, adapting to and mitigating climate change

The quality and quantity of Earth observations continue to mount rapidly. In addition to the ongoing launch of new remote-sensing satellites, increasingly sophisticated in-situ instruments located on the ground, on balloons and airplanes, and in rivers, lakes and oceans, are generating increasingly comprehensive, near-real time observations.

GROUP ON EARTH OBSERVATIONS

The Group on Earth Observations (or GEO) coordinates international efforts to build a Global Earth Observation System of Systems (GEOSS). It links existing and planned Earth observation systems and supports the development of new ones in cases of perceived gaps in the supply of environment-related information. It aims to construct a global public infrastructure for Earth observations consisting in a flexible and distributed network of systems and content providers.

Concept

Common Earth observation instruments include ocean buoys, meteorological stations and balloons, seismic and Global Positioning System (GPS) stations, remote-sensing satellites, computerized forecasting models and early warning systems. These instruments are used to measure and monitor specific aspects of Earth's physical, chemical and biological systems.

To be useful, the raw data collected must be processed, archived, interpreted, and made available via easy-to-use channels in the form of information comprehensible not only by remote sensing experts. Earth observations are vital for policymaking and assessment in many fields.

GEO focuses on facilitating access to Earth observation data for nine priority areas: natural and human-induced disasters, environmental sources of health hazards, energy management, climate change and its impacts, freshwater resources, weather forecasting, ecosystem management, sustainable agriculture, and biodiversity conservation.

History and Structure

GEO was established in February 2005 by the Third Earth Observation Summit in Brussels at the end of a process that started in 2003 with the First Earth Observation Summit in Washington, DC. It was launched in response to calls for action by the 2002 World Summit on Sustainable Development and the Group of Eight (G8) leading industrialized countries. These high-level meetings recognized that international collaboration is essential for exploiting the growing potential of Earth observations to support decision making in an increasingly complex and environmentally stressed world.

GEO is a voluntary partnership of governments and international organizations. It provides a framework within which these partners can develop new projects and coordinate their strategies and investments. As of January 2016, GEO's membership includes 102 governments including the European Commission. In addition, 92 intergovernmental, international and regional organizations with a mandate in Earth observation or related issues have been recognized as participating organizations. Each member and participating organization is represented by a principal and a principal alternate. Members make financial contributions to GEO on a voluntary basis.

GEO is constructing GEOSS on the basis of a 10-Year Strategic Plan from 2016 to 2025. The plan defines a vision statement for GEOSS, its purpose and scope, expected benefits, eight "Societal Benefit Areas" (disaster resilience public health surveillance, energy and mineral resources management, sustainable urban development, water resources management, biodiversity and ecosystem sustainability, food security and sustainable agriculture and infrastructure and transport management - with climate as a cross cutting issue), technical and capacity-building priorities, and the GEO governance structure.

GEO is governed by a plenary consisting of all members and participating Organizations. GEO meets in plenary at least once a year at the level of senior officials and periodically at the ministerial level. Members make decisions at the plenary by consensus.

GLOBAL EARTH OBSERVATION SYSTEM OF SYSTEMS

The Global Earth Observation System of Systems (GEOSS) is being built by the Group on Earth Observations (GEO) on the basis of a 10-Year Implementation Plan running from 2005 to 2015. GEOSS seeks to connect the producers of environmental data and decision-support tools with the end users of these products, with the aim of enhancing the relevance of Earth observations to global issues. GEOSS aims to produce a global public infrastructure that generates comprehensive, near-real-time environmental data, information and analyses for a wide range of users. The Secretariat Director of Geoss is Barbara Ryan.

Earth Observation Systems

Earth observation systems consist of instruments and models designed to measure, monitor and predict the physical, chemical and biological aspects of the Earth system. Buoys floating in the oceans monitor temperature and salinity; meteorological stations and balloons record air quality and rainwater trends; sonar and radar systems estimate fish and bird populations; seismic and Global Positioning System (GPS) stations record movements in the Earth's crust and interior; some 60-plus high-tech environmental satellites scan the planet from space; powerful computerized models generate simulations and forecasts; and early warning systems issue alerts to vulnerable populations.

These various systems have typically operated in isolation from one another. In recent years, however, sophisticated new technologies for gathering vast quantities of near-real-time and high-resolution Earth observation data have become operational. At the same time, improved forecasting models and decision-support tools are increasingly allowing decision makers and other users of Earth observations to fully exploit this widening stream of information.

With investments in Earth observations now reaching a critical mass, it has become possible to link diverse observing systems together to paint a full picture of the Earth's condition. Because the costs and logistics of expanding Earth observations are daunting for any single nation, linking systems together through international cooperation also offers cost savings.

Implementation

As a networked system, GEOSS is owned by all of the GEO Members and Participating Organizations. Partners maintain full control of the components and activities that they contribute to the system of systems. Implementation is being pursued through a Work Plan consisting of over 70 tasks. Each task supports one of the nine societal-benefit or four transverse areas and is carried out by interested Members and Participating Organizations. Governments and organizations have also advanced GEOSS by contributing a variety of

"Early Achievements"; these "First 100 Steps to GEOSS" were presented to the 2007 Cape Town Ministerial Summit.

Interlinking observation systems requires common standards for architecture and data sharing. The architecture of an Earth observation system refers to the way in which its components are designed so that they function as a whole. Each GEOSS component must be included in the GEOSS registry and configured so that it can communicate with the other participating systems. In addition, each contributor to GEOSS must subscribe to the GEO data-sharing principles, which aim to ensure the full and open exchange of data, metadata and products. These issues are fundamental to the successful operation of GEOSS.

GEOSS will disseminate information and analyses directly to users. GEO is developing the GEOPortal as a single Internet gateway to the data produced by GEOSS. The purpose of GEOPortal is to make it easier to integrate diverse data sets, identify relevant data and portals of contributing systems, and access models and other decision-support tools. For users without good access to high-speed internet, GEO has established GEONETCast, a system of four communications satellites that transmit data to low-cost receiving stations maintained by the users.

At present, GEONETCast seems still in its infancy, yet some tools have already been worked out. The GEONETCast toolbox has been made available and contains tools to access some radar altimetry, vegetation, satellite prediction and maritime information. Other useful information available trough GEONETCast is vegetation and desert locust information provided under the DevCoCast project, which is a subproject of GEONETCast.

User Groups

The growing demand for Earth observation data and information is the driving force behind GEOSS. The GEOSS Implementation Plan identifies nine distinct groups of users and uses, which it calls "Societal Benefit Areas". The nine areas are disasters, health, energy, climate, water, weather, ecosystems, agriculture and biodiversity. Current and potential users include decision makers in the public and private sectors, resource managers, planners, emergency responders and scientists.

Related Initiatives

GEOSS can be characterized as a contribution towards the establishment of a spatial data infrastructure. It is one of three related initiatives that are the subject of the GIGAS (GEOSS, INSPIRE and GMES an Action in Support) harmonization project under the auspices of the EU 7th Framework Programme.

EARTH OBSERVATION SATELLITES

Earth observation satellites are satellites specifically designed for Earth observation from orbit, similar to spy satellites but intended for non-military uses such as environmental monitoring, meteorology, map making etc.

Most Earth observation satellites carry instruments that should be operated at a relatively low altitude. Altitudes below 500-600 kilometers are in general avoided, though, because of the significant air-drag at such low altitudes making frequent orbit reboost maneuvres necessary. The Earth observation satellites ERS-1, ERS-2 and Envisat of European Space Agency as well as the MetOp spacecraft of EUMETSAT are all operated at altitudes of about 800 km. The Proba-1, Proba-2 and SMOS spacecraft of European Space Agency are observing the Earth from an altitude of about 700 km. The Earth observation satellites of UAE, DubaiSat-1 & DubaiSat-2 are also placed in Low Earth Orbits (LEO) orbits and providing satellite imagery of various parts of the Earth.

To get (nearly) global coverage with a low orbit it must be a polar orbit or nearly so. A low orbit will have an orbital period of roughly 100 minutes and the Earth will rotate around its polar axis with about 25 deg between successive orbits, with the result that the ground track is shifted towards west with these 25 deg in longitude. Most are in sun-synchronous orbits.

Spacecraft carrying instruments for which an altitude of 36000 km is suitable sometimes use a geostationary orbit. Such an orbit allows uninterrupted coverage of more than 1/3 of the Earth. Three geostationary spacecraft at longitudes separated with 120 deg can cover the whole Earth except the extreme polar regions. This type of orbit is mainly used for meteorological satellites.

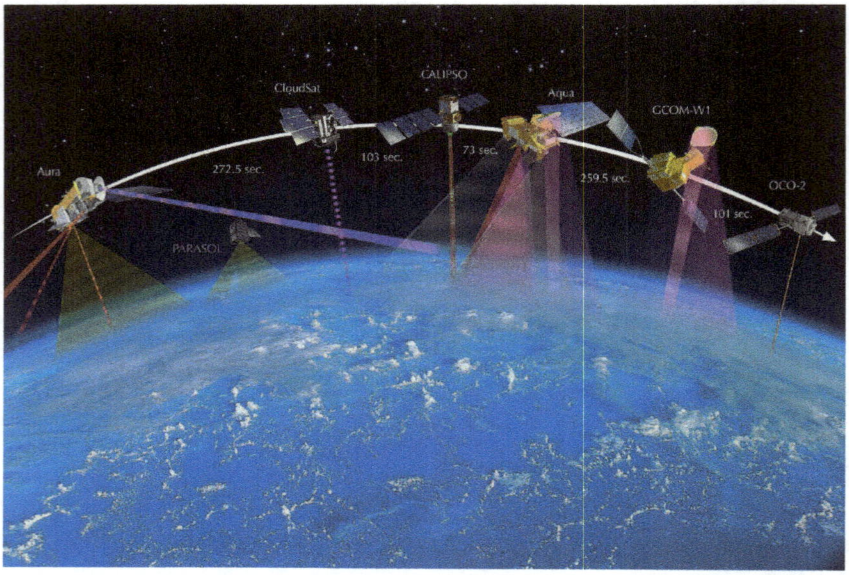

Six Earth observation satellites comprising the A-train satellite constellation as of 2014.

Weather

A weather satellite is a type of satellite that is primarily used to monitor the weather and climate of the Earth. These meteorological satellites, however, see more than clouds and cloud systems. City lights, fires, effects of pollution, auroras, sand and dust storms, snow cover, ice mapping, boundaries of ocean currents, energy flows, etc., are other types of environmental information collected using weather satellites.

GOES-8, a United States weather satellite.

Weather satellite images helped in monitoring the volcanic ash cloud from Mount St. Helens and activity from other volcanoes such as Mount Etna. Smoke from fires in the western United States such as Colorado and Utah have also been monitored.

The El Niño Southern Oscillation and its effects on weather are monitored daily from satellite images. The Antarctic ozone hole is mapped from weather satellite data. Collectively, weather satellites flown by the U.S., Europe, India, China, Russia, and Japan provide nearly continuous observations for a global weather watch, used via visible light and infrared rays of the electromagnetic spectrum. /

Environmental Monitoring

Other environmental satellites can assist environmental monitoring by detecting changes in the Earth's vegetation, atmospheric trace gas content, sea state, ocean color, and ice fields. By monitoring vegetation changes over time, droughts can be monitored by comparing the current vegetation state to its long term average. For example, the 2002 oil spill off the northwest coast of Spain was watched carefully by the European ENVISAT, which, though not a weather satellite, flies an instrument (ASAR) which can see changes in the sea surface. Anthropogenic emissions can be monitored by evaluating data of tropospheric NO_2 and SO_2.

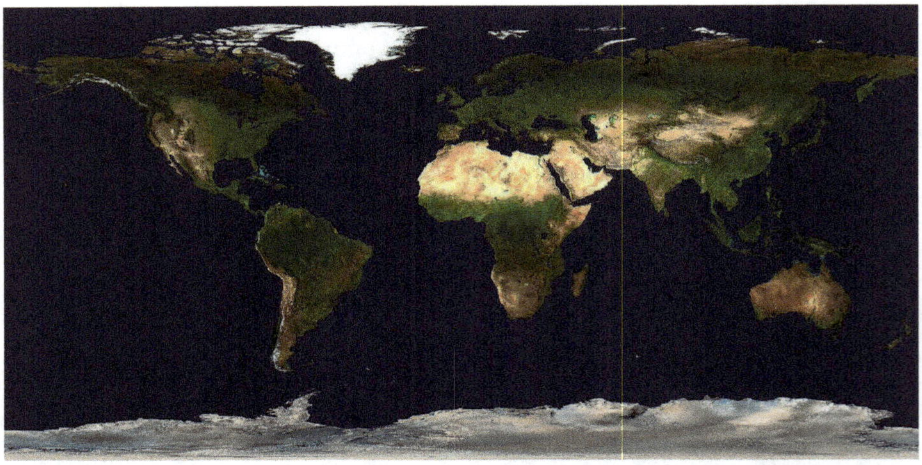

A composite satellite image of the earth, showing its entire surface in Plate carrée projection.

These types of satellites are almost always in Sun synchronous and "frozen" orbits. The Sun synchronous orbit is in general sufficiently close to polar to get the desired global coverage while the relatively constant geometry to the Sun mostly is an advantage for the instruments. The "frozen" orbit is selected as this is the closest to a circular orbit that is possible in the gravitational field of the Earth.

Mapping

Terrain can be mapped from space with the use of satellites, such as Radarsat-1 and TerraSAR-X.

Weather Satellite

The weather satellite is a type of satellite that is primarily used to monitor the weather and climate of the Earth. Satellites can be polar orbiting, covering the entire Earth asynchronously, or geostationary, hovering over the same spot on the equator.

Meteorological satellites see more than clouds and cloud systems. City lights, fires, effects of pollution, auroras, sand and dust storms, snow cover, ice mapping, boundaries of ocean currents, energy flows, etc. Other types of environmental information are collected using weather satellites. Weather satellite images helped in monitoring the volcanic ash cloud from Mount St. Helens and activity from other volcanoes such as Mount Etna. Smoke from fires in the western United States such as Colorado and Utah have also been monitored.

Other environmental satellites can detect changes in the Earth's vegetation, sea state, ocean color, and ice fields. For example, the 2002 Prestige oil spill off the northwest coast of Spain was watched carefully by the European ENVISAT, which, though not a weather satellite, flies an instrument (ASAR) which can see changes in the sea surface.

El Niño and its effects on weather are monitored daily from satellite images. The Antarctic ozone hole is mapped from weather satellite data. Collectively, weather satellites flown by the U.S., Europe, India, China, Russia, and Japan provide nearly continuous observations for a global weather watch.

History

As early as 1946, the idea of cameras in orbit to observe the weather was being developed. This was due to sparse data observation coverage and the expense of using cloud cameras on rockets. By 1958, the early prototypes for TIROS and Vanguard (developed by the Army Signal Corps) were created. The first weather satellite, Vanguard 2, was launched on February 17, 1959. It was designed to measure cloud cover and resistance, but a poor axis of rotation and its elliptical orbit kept it from collecting a notable amount of useful data. The Explorer VI and VII satellites also contained weather-related experiments.

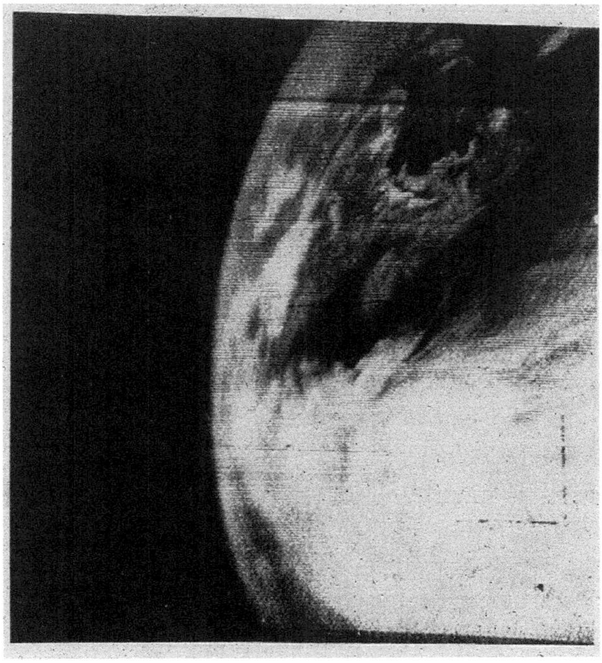

The first television image of Earth from space from the TIROS-1 weather satellite.

The first weather satellite to be considered a success was TIROS-1, launched by NASA on April 1, 1960. TIROS operated for 78 days and proved to be much more successful than Vanguard 2. TIROS paved the way for the Nimbus program, whose technology and findings are the heritage of most of the Earth-observing satellites NASA and NOAA have launched since then. Beginning with the Nimbus 3 satellite in 1969, temperature information through the tropospheric column began to be retrieved by satellites from the eastern Atlantic and most of the Pacific Ocean, which led to significant improvements to weather forecasts.

The ESSA and NOAA polar orbiting satellites followed suit from the late 1960s onward. Geostationary satellites followed, beginning with the ATS and SMS series in the late 1960s and early 1970s, then continuing with the GOES series from the 1970s onward. Polar orbiting satellites such as QuikScat and TRMM began to relay wind information near the ocean's surface starting in the late 1970s, with microwave imagery which resembled radar displays, which significantly improved the diagnoses of tropical cyclone strength, intensification, and location during the 2000s and 2010s.

Observation

Observation is typically made via different 'channels' of the Electromagnetic spectrum, in particular, the Visible and Infrared portions.

Some of these channels include

- *Visible and Near Infrared:* 0.6 μm – 1.6 μm – For recording cloud cover during the day
- *Infrared:* 3.9 μm – 7.3 μm (Water Vapor), 8.7 μm, – 13.4 μm (Thermal imaging)

Visible Spectrum

Visible-light images from weather satellites during local daylight hours are easy to interpret even by the average person; clouds, cloud systems such as fronts and tropical storms, lakes, forests, mountains, snow ice, fires, and pollution such as smoke, smog, dust and haze are readily apparent. Even wind can be determined by cloud patterns, alignments and movement from successive photos.

Infrared Spectrum

The thermal or infrared images recorded by sensors called scanning radiometers enable a trained analyst to determine cloud heights and types, to calculate land and surface water temperatures, and to locate ocean surface features. Infrared satellite imagery can be used effectively for tropical cyclones with a visible eye pattern, using the Dvorak technique, where the difference between the temperature of the warm eye and the surrounding cold cloud tops can be used to determine its intensity (colder cloud tops generally indicate a more intense storm). Infrared pictures depict ocean eddies or vortices and map currents such as the Gulf Stream which are valuable to the shipping industry. Fishermen and farmers are interested in knowing land and water temperatures to protect their crops against frost or increase their catch from the sea. Even El Niño phenomena can be spotted. Using color-digitized techniques, the gray shaded thermal images can be converted to color for easier identification of desired information.

Types

Each meteorological satellite is designed to use one of two different classes of orbit: geostationary and polar orbiting.

The geostationary Himawari 8 satellite's first true-color composite PNG image

Geostationary

Geostationary weather satellites orbit the Earth above the equator at altitudes of 35,880 km (22,300 miles). Because of this orbit, they remain stationary with respect to the rotating Earth and thus can record or transmit images of the entire hemisphere below continuously with their visible-light and infrared sensors. The news media use the geostationary photos in their daily weather presentation as single images or made into movie loops. These are also available on the city forecast pages of noaa.gov (example Dallas, TX).

Several geostationary meteorological spacecraft are in operation. The United States has three in operation; GOES-12, GOES-13, and GOES-15. GOES-12, previously designated GOES-East and now used for South America, is located at 60 degrees west. GOES-13 took over the role of GOES-East on April 14, 2010 and is located at 75 degrees west. GOES-11 was GOES-West over the eastern Pacific Ocean until it was decommissioned December 2011 and replaced by GOES-15. Russia's new-generation weather satellite Elektro-L No.1 operates at 76°E over the Indian Ocean. The Japanese have the MTSAT-2 located over the mid Pacific at 145°E and the Himawari 8 at 140°E. The Europeans have four in operation, Meteosat-8 (3.5°W) and Meteosat-9 (0°) over the Atlantic Ocean and have Meteosat-6 (63°E) and Meteosat-7 (57.5°E) over the Indian Ocean. China currently has three Fengyun (风云) geostationary satellites (FY-2E at 86.5°E, FY-2F at 123.5°E, and FY-2G at 105°E) operated. India also operates geostationary satellites called INSAT which carry instruments for meteorological purposes.

Polar Orbiting

Polar orbiting weather satellites circle the Earth at a typical altitude of 850 km (530 miles) in a north to south (or vice versa) path, passing over the poles in their continuous flight. Polar satellites are in sun-synchronous orbits, which means they are able to observe any place on Earth and will view every location twice each day with the same general lighting conditions due to the near-constant local solar time. Polar orbiting weather satellites offer a much better resolution than their geostationary counterparts due their closeness to the Earth.

Computer controlled motorized parabolic dish antenna for tracking LEO weather satellites.

The United States has the NOAA series of polar orbiting meteorological satellites, presently NOAA 17 and NOAA 18 as primary spacecraft, NOAA 15 and NOAA 16 as secondary spacecraft, NOAA 14 in standby, and NOAA 12. Europe has the Metop-A and Metop-B satellites operated by EUMETSAT. Russia has the Meteor and RESURS series of satellites. China has FY-3A, 3B and 3C. India has polar orbiting satellites as well.

DMSP

The United States Department of Defense's Meteorological Satellite (DMSP) can "see" the best of all weather vehicles with its ability to detect objects almost as 'small' as a huge oil tanker. In addition, of all the weather satellites in orbit, only DMSP can "see" at night in the visual. Some of the most spectacular photos have been recorded by the night visual sensor; city lights, volcanoes, fires, lightning, meteors, oil field burn-offs, as well as the Aurora Borealis and Aurora Australis have been captured by this 450-mile-high space vehicle's low moonlight sensor.

Turnstile antenna for reception of 137 MHz LEO weather satellite transmissions

At the same time, energy use and city growth can be monitored since both major and even minor cities, as well as highway lights, are conspicuous. This informs astronomers of light pollution. The New York City Blackout of 1977 was captured by one of the night orbiter DMSP space vehicles.

In addition to monitoring city lights, these photos are a life saving asset in the detection and monitoring of fires. Not only do the satellites see the fires visually day and night, but the thermal and infrared scanners on board these weather satellites detect potential fire sources below the surface of the Earth where smoldering occurs. Once the fire is detected, the same weather satellites provide vital information about wind that could fan or spread the fires. These same cloud photos from space tell the firefighter when it will rain.

Some of the most dramatic photos showed the 600 Kuwaiti oil fires that the fleeing Army of Iraq started on February 23, 1991. The night photos showed huge flashes, far outstripping the glow of large populated areas. The fires consumed millions of gallons of oil; the last was doused on November 6, 1991.

Uses

Snowfield monitoring, especially in the Sierra Nevada, can be helpful to the hydrologist keeping track of available snowpack for runoff vital to the watersheds of the western United States. This information is gleaned from existing satellites of all agencies of the U.S. government (in addition to local, on-the-ground measurements). Ice floes, packs and bergs can also be located and tracked from weather space craft.

Even pollution whether it is nature-made or man-made can be pinpointed. The visual and infrared photos show effects of pollution from their respective areas over the entire earth. Aircraft and rocket pollution, as well as condensation trails, can also be spotted. The ocean current and low level wind information gleaned from the space photos can help predict oceanic oil spill coverage and movement. Almost every summer, sand and dust from the Sahara Desert in Africa drifts across the equatorial regions of the Atlantic Ocean. GOES-EAST photos enable meteorologists to observe, track and forecast this sand cloud. In addition to reducing visibilities and causing respiratory problems, sand clouds suppress hurricane formation by modifying the solar radiation balance of the tropics. Other dust storms in Asia and mainland China are common and easy to spot and monitor, with recent examples of dust moving across the Pacific Ocean and reaching North America.

In remote areas of the world with few local observers, fires could rage out of control for days or even weeks and consume millions of acres before authorities are alerted. Weather satellites can be a tremendous asset in such situations. Nighttime photos also show the burn-off in gas and oil fields. Atmospheric temperature and moisture profiles have been taken by weather satellites since 1969.

TerraSAR-X

TerraSAR-X, a radar Earth observation satellite, is a joint venture being carried out under a public-private-partnership between the German Aerospace Center (DLR) and EADS Astrium. The exclusive commercial exploitation rights are held by the geo-information service provider Astrium. TerraSAR-X was launched on June 15, 2007 and has been in operational service since January 2008. With its twin satellite TanDEM-X, launched June 21, 2010, TerraSAR-X acquires the data basis for the WorldDEM, the worldwide and homogeneous DEM available from 2014.

Satellite and Mission

Radar Image

With its phased array synthetic aperture radar (SAR) antenna (X-band wavelength 31 mm, frequency 9.6 GHz), TerraSAR-X acquires new high-quality radar images of the entire planet whilst circling Earth in a polar orbit at 514 km altitude. The orbit is selected such that the satellite flies in a sun-synchronous dusk-dawn orbit, which means that it moves along the day-night boundary of the Earth and always presents the same face to the sun, ensuring an optimum energy supply via the solar cells. TerraSAR-X is designed to carry out its task for five years, independent of weather conditions and illumination, and reliably provides radar images with a resolution of up to 1m.

Features of TerraSAR-X:

- resolution of up to 1 m,
- excellent radiometric accuracy,
- geometric accuracy unrivalled by any other commercial spaceborne sensor,
- quick site access time of 2.5 days max. (2 days at 95% probability) to any point on Earth,
- unique agility (rapid switches between imaging modes and polarisations)

TerraSAR-X Imaging Modes

TerraSAR-X acquires radar data in the following three main imaging modes:

- SpotLight: up to 1 m resolution, scene size 10 km (width) × 5 km (length);
- StripMap: up to 3 m resolution, scene size 30 km (width) × 50 km (length);
- ScanSAR: up to 16 m resolution, scene size 100 km (width) × 150 km (length);

In addition, the unique design of TerraSAR-X›s SAR antenna allows a variety of polarimetric combinations: single or dual polarization and even full polarimetric data takes, are possible.

Depending on the desired application, one of four different product types (processing levels) can be selected

- Single Look Slant Range Complex (SSC)
- Multi Look Ground Range Detected (MGD)
- Geocoded Ellipsoid Corrected (GEC)
- Enhanced Ellipsoid Corrected (EEC)

TanDEM-X and WorldDEM Akida

TanDEM-X (TerraSAR-X add-on for Digital Elevation Measurements) is a second, very similar spacecraft launched successfully on June 21, 2010 from Baikonur Cosmodrome in Kazakhstan. Since October 2010, TerraSAR-X and TanDEM-X fly in a close formation at distances of only a few hundred metres and record data synchronously. This unique twin satellite constellation will allow the generation of WorldDEM, the global digital elevation models (DEMs). With an unprecedented accuracy, coverage and quality – WorldDEM is a consistent DEM of the Earth's land surface is envisaged to be acquired and generated within three years after launch. Available from 2014, WorldDEM is to feature a vertical accuracy of 2m (relative) and 10m (absolute), within a horizontal raster of approximately 12x12 square meters, slightly varying depending on the geographic latitude.

Advantages of Radar

Radar stood originally for *Radio Detection and Ranging* and contains traditionally:

- Range finding (EDM) by means of the time a reflected signal needs to return;
- Direction measurement over the adjustment of the antenna, and recently;
- also different analysis such as SAR, polarization, Interferometry etc.

Satellite radar entered use more than fifteen years after optical camera systems did. The resolution (detail sharpness) is in principle lower, but radar can gather information at any time of the day or night and independent of cloud cover. This contributes significantly to the reliability of the system, a much-desired property.

Early radar satellite techniques were e.g. the Altimetrie (leveling over the sea), NASA's SEASAT (launched in 1978), regulation of waves/wind or soil data. Satellites like GRACE can measure for instance the speed of other satellites to mm/s exactly, and the slow deformation of volcanoes. The military has used radar since the late 1930s and radar satellites at least since 1978.

Innovations with TerraSAR X

TerraSAR X will exhibit some technical-industrial novelties. One of these innovations is a kind of zoom shot, with the resolution and scanning field vice versa changeable in a 1:10 relationship, either a larger area to grasp or a small area with the highest possible resolution.

Furthermore, the antenna can be aligned by electronics within an angle range so that the point of view is adjustable. Earlier radar satellites could radiate the antenna only in one direction.

Scanning and Trajectory

With the adjustable angle radar sensor - along with other course refinements (precession by the earth flattening) - any place on earth can be observed preferentially within 1–3 days.

For a specific point on the Earth's equator, TerraSAR X has a revisit cycle of 11 days. The revisit time decreases towards the poles, e.g. northern Europe has a revisit time of typically 3–4 days.

Ground Segment

The ground operating mechanism and controls for the TerraSAR X is developed by the DLR in Oberpfaffenhofen. It consists of Mission Operating Equipment, the Payload Ground Segment and the Instrument Operation and Calibration Segment. At the base of the ground segment lies the German Space Operation Center (GSOC), the German Remote Sensing Datum Center (DFD) as well as Institutes for Methodology of Remote Sensing (MF) and the Institute for High-Frequency Engineering and Radar Systems (HR) which are all part of the DLR.

Applications

Applications of the high-resolution TerraSAR-X radar imagery include:

- Topographic Mapping: 2D and 3D, in scales down to 1:25,000, map updates
- Surface Movement: Based on timeseries acquired by TerraSAR-X over the same area surface displacements caused by subsurface mining, oil-/gas extraction, infrastructure construction, excavations, or underground engineering can be visualised.
- Change Detection: for the monitoring of large-scale construction projects, infrastructure networks, monitoring and documentation of changes and developments
- Land Cover and Land Use Mapping: accurate and up-to-date land cover / land use informations, also from places, where it is difficult to get informations with using other technologies because of permanent cloud cover
- Defence and Security Applications: Applications include effective mission planning, the quick assessment of natural or manmade disasters or border control through detection of paths (changes), fences and moving objects
- Rapid Emergency Response: due to its rapid revisit time TerraSAR-X is a reliable source of information in case of natural or man-made disasters (e.g. earth quakes, floods, military conflicts etc.) providing reliable information for disaster management and response allowing the recognition and assessment of damages to populated areas and traffic infrastructure, the identification of focus areas, and an efficient coordination of rescue actions.
- Environmental applications: e.g. forest monitoring, flood monitoring, water quality applications
- Further applications currently under evaluation: Traffic Monitoring, Maritime applications, vegetation monitoring

Scientific use of TerraSAR-X Data

The scientific use of the TerraSAR-X data will be coordinated through the TerraSAR-X Science Service System by the DLR. The new-quality data records, as provided by TerraSAR-X, will offer a vast amount of new research incentives, for instance in ecology, geology, hydrology and oceanography. The smallest movements of the Earth's surface (plate tectonics, volcanism, earthquake) are further scientific fields of application.

Commercial Use of TerraSAR-X Data

In order to ensure the commercial success of the mission, EADS Astrium founded its 100% subsidiary Infoterra in 2001; the company being responsible for establishing a commercial market for TerraSAR-X data as well as TerraSAR-X-based geo-information products and services.

REFERENCES

1. "International Journal of Applied Earth Observation and Geoinformation -- Aims and Scope". Elsevier. Retrieved 20 July 2012.
2. "DubaiSat-2, Earth Observation Satellite of UAE". Mohammed Bin Rashid Space Centre.
3. "DubaiSat-1, Earth Observation Satellite of UAE". Mohammed Bin Rashid Space Centre.
4. NESDIS. Satellites. Retrieved on 2008-07-04.
5. NOAA. NOAA Satellites, Scientists Monitor Mt. St. Helens for Possible Eruption. Retrieved on 2008-07-04.
6. NASA. Drought. Retrieved on 2008-07-04.
7. Grunsky, E.C. The use of multi-beam RADARSAT-1 satellite imagery for terrain mapping. Retrieved on 2008-07-04.
8. NESDIS. Satellites. Retrieved on July 4, 2008.
9. NOAA. NOAA Satellites, Scientists Monitor Mt. St. Helens for Possible Eruption. Retrieved on July 4, 2008.
10. Janice Hill (1991). Weather From Above: America's Meteorological Satellites. Smithsonian Institution. pp. 4–7. ISBN 0-87474-394-X.
11. "VANGUARD - A HISTORY, CHAPTER 12, SUCCESS - AND AFTER". NASA. Archived from the original on 2008-05-09.
12. "U.S. Launches Camera Weather Satellite". The Fresno Bee. AP and UPI. April 1, 1960. pp. 1a, 4a.
13. National Environmental Satellite Center (January 1970). "SIRS and the Improved Marine Weather Forecast". Mariners Weather Log. Environmental Science Services Administration. 14 (1): 12–15.
14. EUMETSAT – MSG Spectrum (PDF)
15. EUMETSAT – MFG Payload
16. F. Hasler, K. Palaniappan, C. Kambhammetu, P. Black, E. Uhlhorn, and D. Chesters. High-Resolution Wind Fields within the Inner Core and Eye of a Mature Tropical Cyclone from GOES 1-min Images. Retrieved on 2008-07-04.
17. Chris Landsea. Subject: H1) What is the Dvorak technique and how is it used? Retrieved on January 3, 2009.
18. http://forecast.weather.gov/MapClick.php?lat=32.81036168486901 5&lon=-96.73736572265625&site=fwd&smap=1&marine=1&unit=0&lg=en
19. "GOES 12 Spacecraft Status Summary". NOAA Satellite and Information Service. Retrieved December 13, 2010.
20. "GOES 13 Spacecraft Status Summary". NOAA Satellite and Information Service. Retrieved February 15, 2012.
21. "GOES 15 Spacecraft Status Summary". NOAA Satellite and Information Service. Retrieved February 15, 2012.

22. National Satellite Meteorological Center of CMA (in Chinese) http://www.nsmc.cma.gov.cn/NSMC/Channels/100003.html. Missing or empty |title= (help)
23. Ann K. Cook (July 1969). "The Breakthrough Team" (PDF). ESSA World. Environmental Satellite Services Administration: 28–31. Retrieved 2012-04-21.
24. "TERRA SAR X Satellite details 2007-026A NORAD 31698". N2YO. 25 January 2015. Retrieved 25 January 2015.
25. StripMap & ScanSAR: acquisition length extendable to up to 1,650 km.
26. DLR - Blogs - The satellites have 'eye contact'
27. GIM International: Weber, Marco; Koudogbo, Fifamè,January 2009, TerraSAR-X 1m Spaceborne Radar - Use, Features, Products and TanDEM-X.
28. (Jensen, J. R. 2007. Remote Sensing of the Environment: An Earth Resource Perspective)
29. GeoBerichte 14, Landesamt für Bergbau, Energie und Geologie in Niedersachsen:Schrage, Thomas;Jacob, Philipp , June 2009, Flächenverbrauch und Bodenversigelung in Niedersachsen.
30. GIM International: Balz, Timo; Scheuchl, Bernd;Li, Deren, October 2008, The Sichuan Earthquake(1)-Satellite Imagery for Rapid Response.
31. GIM International: Shao, Yun; Scheuchl, Bernd, November 2008, The Sichuan Earthquake (2)- Spaceborne SAR in Earthquake Response.
32. GIM International: Koudogbo, Fifamè; Müller, Marc; Scheuchl, Bernd, December 2008, The Sichuan Earthquake (3)- Satellite-based Global Flood Response.
33. TerraSAR-X Science Service System

CHAPTER 4

Global Positioning System

Chapter Contents

The global positioning system is a satellite-based navigation system consisting of a network of 24 orbiting satellites that are eleven thousand nautical miles in space and in six different orbital paths. The satellites are constantly moving, making two complete orbits around the Earth in just under 24 hours. If you do the math, that's about 2.6 kilometers per second. That's really moving!

INTRODUCTION

The Global Positioning System (GPS), also known as Navstar GPS or simply Navstar, is a global navigation satellite system (GNSS) that provides geolocation and time information to a GPS receiver in all weather conditions, anywhere on or near the Earth where there is an unobstructed line of sight to four or more GPS satellites. The GPS system operates independently of any telephonic or internet reception, though these technologies can enhance the usefulness of the GPS positioning information. The GPS system provides critical positioning capabilities to military, civil, and commercial users around the world. The

United States government created the system, maintains it, and makes it freely accessible to anyone with a GPS receiver. However, the US government can selectively deny access to the system, as happened to the Indian military in 1999 during the Kargil War.

The GPS project was launched in the United States in 1973 to overcome the limitations of previous navigation systems, integrating ideas from several predecessors, including a number of classified engineering design studies from the 1960s. The U.S. Department of Defense (DoD) developed the system, which originally used 24 satellites. It became fully operational in 1995. Roger L. Easton of the Naval Research Laboratory, Ivan A. Getting of The Aerospace Corporation, and Bradford Parkinson of the Applied Physics Laboratory are credited with inventing it.

Advances in technology and new demands on the existing system have now led to efforts to modernize the GPS and implement the next generation of GPS Block IIIA satellites and Next Generation Operational Control System (OCX). Announcements from Vice President Al Gore and the White House in 1998 initiated these changes. In 2000, the U.S. Congress authorized the modernization effort, GPS III.

In addition to GPS, other systems are in use or under development, mainly because of a potential denial of access and potential monitoring by the US government. The Russian Global Navigation Satellite System (GLONASS) was developed contemporaneously with GPS, but suffered from incomplete coverage of the globe until the mid-2000s.GLONASS can be added to GPS devices which makes more satellites available and enabling positions to be fixed more quickly and accurately, to within 2 meters There are also the European Union Galileo positioning system, China's BeiDou Navigation Satellite System, the Japanese Quasi-Zenith Satellite System, and India's Indian Regional Navigation Satellite System (NAVIC).

Artist's conception of GPS Block II-F satellite in Earth orbit.

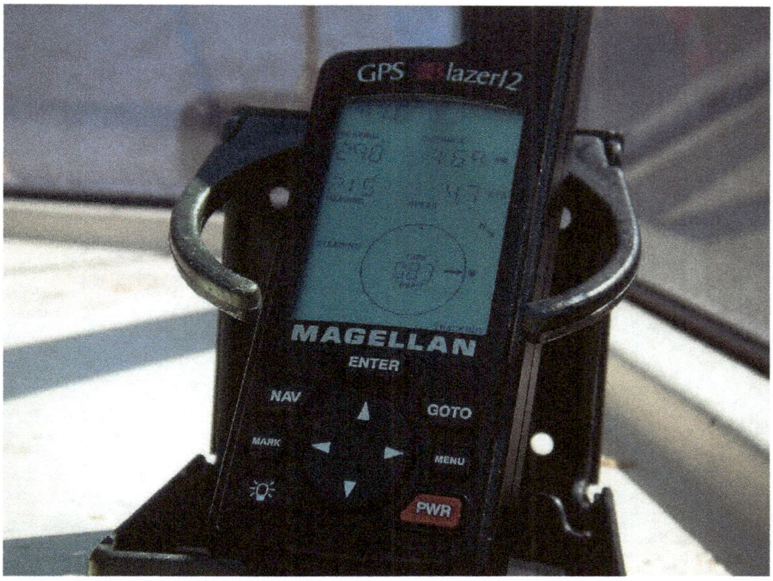

Civilian GPS receivers ("GPS navigation device") in a marine application.

Automotive navigation system in a taxicab.

A U.S. Air Force Senior Airman runs through a checklist during Global Positioning System satellite operations.

HISTORY

The design of GPS is based partly on similar ground-based radio-navigation systems, such as LORAN and the Decca Navigator, developed in the early 1940s and used by the British Royal Navy during World War II.

In 1956, the German-American physicist Friedwardt Winterberg proposed a test of general relativity — detecting time slowing in a strong gravitational field using accurate atomic clocks placed in orbit inside artificial satellites.

Special and general relativity predict that the clocks on the GPS satellites would be seen by the Earth's observers to run 38 microseconds faster per day than the clocks on the Earth. The GPS calculated positions would quickly drift into error, accumulating to 10 kilometers per day. The relativistic time effect of the GPS clocks running faster than the clocks on earth was corrected for in the design of GPS.

Predecessors

The Soviet Union launched the first man-made satellite, Sputnik 1, in 1957. Two American physicists, William Guier and George Weiffenbach, at Johns Hopkins's Applied Physics Laboratory (APL), decided to monitor Sputnik's radio transmissions. Within hours they realized that, because of the Doppler effect, they could pinpoint where the satellite was along its orbit. The Director of the APL gave them access to their UNIVAC to do the heavy calculations required.

The next spring, Frank McClure, the deputy director of the APL, asked Guier and Weiffenbach to investigate the inverse problem — pinpointing the user's location, given that of the satellite. (At the time, the Navy was developing the submarine-launched Polaris missile, which required them to know the submarine's location.) This led them and APL to develop the TRANSIT system. In 1959, ARPA (renamed DARPA in 1972) also played a role in TRANSIT.

Official logo for NAVSTAR GPS

Emblem of the 50th Space Wing

The first satellite navigation system, TRANSIT, used by the United States Navy, was first successfully tested in 1960. It used a constellation of five satellites and could provide a navigational fix approximately once per hour.

In 1967, the U.S. Navy developed the Timation satellite that proved the ability to place accurate clocks in space, a technology required by GPS.

In the 1970s, the ground-based OMEGA navigation system, based on phase comparison of signal transmission from pairs of stations, became the first worldwide radio navigation system. Limitations of these systems drove the need for a more universal navigation solution with greater accuracy.

While there were wide needs for accurate navigation in military and civilian sectors, almost none of those was seen as justification for the billions of dollars it would cost in research, development, deployment, and operation for a constellation of navigation satellites. During the Cold War arms race, the nuclear threat to the existence of the United States was the one need that did justify this cost in the view of the United States Congress. This deterrent effect is why GPS was funded. It is also the reason for the ultra secrecy at that time. The nuclear triad consisted of the United States Navy's submarine-launched ballistic missiles (SLBMs) along with United States Air Force (USAF) strategic bombers and intercontinental ballistic missiles (ICBMs). Considered vital to the nuclear deterrence posture, accurate determination of the SLBM launch position was a force multiplier.

Precise navigation would enable United States ballistic missile submarines to get an accurate fix of their positions before they launched their SLBMs. The USAF, with two thirds of the nuclear triad, also had requirements for a more accurate and reliable navigation system. The Navy and Air Force were developing their own technologies in parallel to solve what was essentially the same problem.

To increase the survivability of ICBMs, there was a proposal to use mobile launch platforms (such as Russian SS-24 and SS-25) and so the need to fix the launch position had similarity to the SLBM situation.

In 1960, the Air Force proposed a radio-navigation system called MOSAIC (MObile System for Accurate ICBM Control) that was essentially a 3-D LORAN. A follow-on study, Project 57, was worked in 1963 and it was "in this study that the GPS concept was born." That same year, the concept was pursued as Project 621B, which had "many of the attributes that you now see in GPS" and promised increased accuracy for Air Force bombers as well as ICBMs.

Updates from the Navy TRANSIT system were too slow for the high speeds of Air Force operation. The Naval Research Laboratory continued advancements with their Timation (Time Navigation) satellites, first launched in 1967, and with the third one in 1974 carrying the first atomic clock into orbit.

Another important predecessor to GPS came from a different branch of the United States military. In 1964, the United States Army orbited its first Sequential Collation of Range (SECOR) satellite used for geodetic surveying. The SECOR system included three ground-based transmitters from known locations that would send signals to the satellite transponder in orbit. A fourth ground-based station, at an undetermined position, could then use those signals to fix its location precisely. The last SECOR satellite was launched in 1969.

Decades later, during the early years of GPS, civilian surveying became one of the first fields to make use of the new technology, because surveyors could reap benefits of signals from the less-than-complete GPS constellation years before it was declared operational. GPS can be thought of as an evolution of the SECOR system where the ground-based transmitters have been migrated into orbit.

Development

With these parallel developments in the 1960s, it was realized that a superior system could be developed by synthesizing the best technologies from 621B, Transit, Timation, and SECOR in a multi-service program.

During Labor Day weekend in 1973, a meeting of about twelve military officers at the Pentagon discussed the creation of a *Defense Navigation Satellite System (DNSS)*. It was at this meeting that the real synthesis that became GPS was created. Later that year, the DNSS program was named *Navstar*, or Navigation System Using Timing and Ranging. With the individual satellites being associated with the name Navstar (as with the predecessors Transit and Timation), a more fully encompassing name was used to identify the constellation of Navstar satellites, *Navstar-GPS*. Ten "Block I" prototype satellites were launched between 1978 and 1985 (an additional unit was destroyed in a launch failure).

After Korean Air Lines Flight 007, a Boeing 747 carrying 269 people, was shot down in 1983 after straying into the USSR's prohibited airspace, in the vicinity of Sakhalin and Moneron Islands, President Ronald Reagan issued a directive making GPS freely available for civilian use, once it was sufficiently developed, as a common good. The first Block II satellite was launched on February 14, 1989, and the 24th satellite was launched in 1994. The GPS program cost at this point, not including the cost of the user equipment, but including the costs of the satellite launches, has been estimated at about USD$5 billion (then-year dollars). Roger L. Easton is widely credited as the primary inventor of GPS.

Initially, the highest quality signal was reserved for military use, and the signal available for civilian use was intentionally degraded (Selective Availability). This changed with President Bill Clinton signing a policy directive in 1996 to turn off Selective Availability in May 2000 to provide the same precision to civilians that was afforded to the military. The directive was proposed by the U.S. Secretary of Defense, William Perry, because of the widespread growth of differential GPS services to improve civilian accuracy and eliminate the U.S. military advantage. Moreover, the U.S. military was actively developing technologies to deny GPS service to potential adversaries on a regional basis.

Since its deployment, the U.S. has implemented several improvements to the GPS service including new signals for civil use and increased accuracy and integrity for all users, all the while maintaining compatibility with existing GPS equipment. Modernization of the satellite system has been an ongoing initiative by the U.S. Department of Defense through a series of satellite acquisitions to meet the growing needs of the military, civilians, and the commercial market.

As of early 2015, high-quality, FAA grade, Standard Positioning Service (SPS) GPS receivers provide horizontal accuracy of better than 3.5 meters, although many factors such as receiver quality and atmospheric issues can affect this accuracy.

GPS is owned and operated by the United States Government as a national resource. The Department of Defense is the steward of GPS. *Interagency GPS Executive Board (IGEB)* oversaw GPS policy matters from 1996 to 2004. After that the National Space-Based Positioning, Navigation and Timing Executive Committee was established by presidential directive in 2004 to advise and coordinate federal departments and agencies on matters concerning the GPS and related systems. The executive committee is chaired jointly by the deputy secretaries of defense and transportation. Its membership includes equivalent-level officials from the departments of state, commerce, and homeland security, the joint chiefs of staff, and NASA. Components of the executive office of the president participate as observers to the executive committee, and the FCC chairman participates as a liaison.

The U.S. Department of Defense is required by law to "maintain a Standard Positioning Service (as defined in the federal radio navigation plan and the standard positioning service signal specification) that will be available on a continuous, worldwide basis," and "develop measures to prevent hostile use of GPS and its augmentations without unduly disrupting or degrading civilian uses."

Timeline and Modernization

Summary of satellites

Block	**Launch period**	**Satellite launches**				**Currently in orbit and healthy**
		Success	**Failure**	**In preparation**	**Planned**	
I	1978–1985	10	1	0	0	0
II	1989–1990	9	0	0	0	0
IIA	1990–1997	19	0	0	0	0
IIR	1997–2004	12	1	0	0	12
IIR-M	2005–2009	8	0	0	0	7
IIF	2010–2016	12	0	0	0	12
IIIA	From 2017	0	0	0	12	0
IIIB	—	0	0	0	8	0
IIIC	—	0	0	0	16	0
Total		70	2	0	36	31

(Last update: March 9, 2016)

8 satellites from Block IIA are placed in reserve
USA-203 from Block IIR-M is unhealthy

In 1972, the USAF Central Inertial Guidance Test Facility (Holloman AFB), conducted developmental flight tests of two prototype GPS receivers over White Sands Missile Range, using ground-based pseudo-satellites.

- In 1978, the first experimental Block-I GPS satellite was launched.
- In 1983, after Soviet interceptor aircraft shot down the civilian airliner KAL 007 that strayed into prohibited airspace because of navigational errors, killing all 269 people on board, U.S. President Ronald Reagan announced that GPS would be made available for civilian uses once it was completed, although it had been previously published [in Navigation magazine] that the CA code (Coarse Acquisition code) would be available to civilian users.
- By 1985, ten more experimental Block-I satellites had been launched to validate the concept.
- Beginning in 1988, Command & Control of these satellites was transitioned from Onizuka AFS, California to the 2nd Satellite Control Squadron (2SCS) located at Falcon Air Force Station in Colorado Springs, Colorado.
- On February 14, 1989, the first modern Block-II satellite was launched.
- The Gulf War from 1990 to 1991 was the first conflict in which the military widely used GPS.
- In 1991, a project to create a miniature GPS receiver successfully ended, replacing the previous 23 kg military receivers with a 1.25 kg handheld receiver.
- In 1992, the 2nd Space Wing, which originally managed the system, was inactivated and replaced by the 50th Space Wing.
- By December 1993, GPS achieved initial operational capability (IOC), indicating a full constellation (24 satellites) was available and providing the Standard Positioning Service (SPS).
- Full Operational Capability (FOC) was declared by Air Force Space Command (AFSPC) in April 1995, signifying full availability of the military's secure Precise Positioning Service (PPS).
- In 1996, recognizing the importance of GPS to civilian users as well as military users, U.S. President Bill Clinton issued a policy directive declaring GPS a dual-use system and establishing an Interagency GPS Executive Board to manage it as a national asset.
- In 1998, United States Vice President Al Gore announced plans to upgrade GPS with two new civilian signals for enhanced user accuracy and reliability, particularly with respect to aviation safety and in 2000 the United States Congress authorized the effort, referring to it as *GPS III*.
- On May 2, 2000 "Selective Availability" was discontinued as a result of the 1996 executive order, allowing users to receive a non-degraded signal globally.
- In 2004, the United States Government signed an agreement with the European Community establishing cooperation related to GPS and Europe's Galileo system.
- In 2004, United States President George W. Bush updated the national policy and replaced the executive board with the National Executive Committee for Space-Based Positioning, Navigation, and Timing.
- November 2004, Qualcomm announced successful tests of assisted GPS for mobile phones.

- In 2005, the first modernized GPS satellite was launched and began transmitting a second civilian signal (L2C) for enhanced user performance.
- On September 14, 2007, the aging mainframe-based Ground Segment Control System was transferred to the new Architecture Evolution Plan.
- On May 19, 2009, the United States Government Accountability Office issued a report warning that some GPS satellites could fail as soon as 2010.
- On May 21, 2009, the Air Force Space Command allayed fears of GPS failure saying "There's only a small risk we will not continue to exceed our performance standard."
- On January 11, 2010, an update of ground control systems caused a software incompatibility with 8000 to 10000 military receivers manufactured by a division of Trimble Navigation Limited of Sunnyvale, Calif.
- On February 25, 2010, the U.S. Air Force awarded the contract to develop the GPS Next Generation Operational Control System (OCX) to improve accuracy and availability of GPS navigation signals, and serve as a critical part of GPS modernization.

Awards

On February 10, 1993, the National Aeronautic Association selected the GPS Team as winners of the 1992 Robert J. Collier Trophy, the nation's most prestigious aviation award. This team combines researchers from the Naval Research Laboratory, the USAF, the Aerospace Corporation, Rockwell International Corporation, and IBM Federal Systems Company. The citation honors them "for the most significant development for safe and efficient navigation and surveillance of air and spacecraft since the introduction of radio navigation 50 years ago."

Two GPS developers received the National Academy of Engineering Charles Stark Draper Prize for 2003:

- Ivan Getting, emeritus president of The Aerospace Corporation and an engineer at the Massachusetts Institute of Technology, established the basis for GPS, improving on the World War II land-based radio system called LORAN (*Lo*ng-range *R*adio *A*id to *N*avigation).
- Bradford Parkinson, professor of aeronautics and astronautics at Stanford University, conceived the present satellite-based system in the early 1960s and developed it in conjunction with the U.S. Air Force. Parkinson served twenty-one years in the Air Force, from 1957 to 1978, and retired with the rank of colonel.

GPS developer Roger L. Easton received the National Medal of Technology on February 13, 2006.

Francis X. Kane (Col. USAF, ret.) was inducted into the U.S. Air Force Space and Missile Pioneers Hall of Fame at Lackland A.F.B., San Antonio, Texas, March 2, 2010 for his role in space technology development and the engineering design concept of GPS conducted as part of Project 621B.

In 1998, GPS technology was inducted into the Space Foundation Space Technology Hall of Fame.

On October 4, 2011, the International Astronautical Federation (IAF) awarded the Global Positioning System (GPS) its 60th Anniversary Award, nominated by IAF member, the American Institute for Aeronautics and Astronautics (AIAA). The IAF Honors and Awards Committee recognized the uniqueness of the GPS program and the exemplary role it has played in building international collaboration for the benefit of humanity.

BASIC CONCEPT OF GPS

Fundamentals

The GPS concept is based on time and the known position of specialized satellites. The satellites carry very stable atomic clocks that are synchronized with one another and to ground clocks. Any drift from true time maintained on the ground is corrected daily. Likewise, the satellite locations are known with great precision. GPS receivers have clocks as well; however, they are usually not synchronized with true time, and are less stable. GPS satellites continuously transmit their current time and position. A GPS receiver monitors multiple satellites and solves equations to determine the precise position of the receiver and its deviation from true time. At a minimum, four satellites must be in view of the receiver for it to compute four unknown quantities (three position coordinates and clock deviation from satellite time).

More Detailed Description

Each GPS satellite continually broadcasts a signal (carrier wave with modulation) that includes:

- A pseudorandom code (sequence of ones and zeros) that is known to the receiver. By time-aligning a receiver-generated version and the receiver-measured version of the code, the time of arrival (TOA) of a defined point in the code sequence, called an epoch, can be found in the receiver clock time scale
- A message that includes the time of transmission (TOT) of the code epoch (in GPS system time scale) and the satellite position at that time

Conceptually, the receiver measures the TOAs (according to its own clock) of four satellite signals. From the TOAs and the TOTs, the receiver forms four time of flight (TOF) values, which are (given the speed of light) approximately equivalent to receiver-satellite range differences. The receiver then computes its three-dimensional position and clock deviation from the four TOFs.

In practice the receiver position (in three dimensional Cartesian coordinates with origin at the Earth's center) and the offset of the receiver clock relative to the GPS time are computed simultaneously, using the navigation equations to process the TOFs.

The receiver's Earth-centered solution location is usually converted to latitude, longitude and height relative to an ellipsoidal Earth model. The height may then be further converted to height relative to the geoid (e.g., EGM96) (essentially, mean sea level). These coordinates may be displayed, e.g., on a moving map display, and/or recorded and/or used by some other system (e.g., a vehicle guidance system).

User-Satellite Geometry

Although usually not formed explicitly in the receiver processing, the conceptual time differences of arrival (TDOAs) define the measurement geometry. Each TDOA corresponds to a hyperboloid of revolution. The line connecting the two satellites involved (and its extensions) forms the axis of the hyperboloid. The receiver is located at the point where three hyperboloids intersect.

It is sometimes incorrectly said that the user location is at the intersection of three spheres. While simpler to visualize, this is only the case if the receiver has a clock synchronized with the satellite clocks (i.e., the receiver measures true ranges to the satellites rather than range differences). There are significant performance benefits to the user carrying a clock synchronized with the satellites. Foremost is that only three satellites are needed to compute a position solution. If this were part of the GPS system concept so that all users needed to carry a synchronized clock, then a smaller number of satellites could be deployed. However, the cost and complexity of the user equipment would increase significantly.

Receiver in Continuous Operation

The description above is representative of a receiver start-up situation. Most receivers have a track algorithm, sometimes called a *tracker*, that combines sets of satellite measurements collected at different times—in effect, taking advantage of the fact that successive receiver positions are usually close to each other. After a set of measurements are processed, the tracker predicts the receiver location corresponding to the next set of satellite measurements. When the new measurements are collected, the receiver uses a weighting scheme to combine the new measurements with the tracker prediction. In general, a tracker can (a) improve receiver position and time accuracy, (b) reject bad measurements, and (c) estimate receiver speed and direction.

The disadvantage of a tracker is that changes in speed or direction can only be computed with a delay, and that derived direction becomes inaccurate when the distance traveled between two position measurements drops below or near the random error of position measurement. GPS units can use measurements of the Doppler shift of the signals received to compute velocity accurately. More advanced navigation systems use additional sensors like a compass or an inertial navigation system to complement GPS.

Non-Navigation Applications

In typical GPS operation as a navigator, four or more satellites must be visible to obtain an accurate result. The solution of the navigation equations gives the position of the receiver along with the difference between the time kept by the receiver's on-board clock and the true time-of-day, thereby eliminating the need for a more precise and possibly impractical receiver based clock. Applications for GPS such as time transfer, traffic signal timing, and synchronization of cell phone base stations, make use of this cheap and highly accurate timing. Some GPS applications use this time for display, or, other than for the basic position calculations, do not use it at all.

Although four satellites are required for normal operation, fewer apply in special cases. If one variable is already known, a receiver can determine its position using only three satellites. For example, a ship or aircraft may have known elevation. Some GPS receivers may use additional clues or assumptions such as reusing the last known altitude, dead reckoning, inertial navigation, or including information from the vehicle computer, to give a (possibly degraded) position when fewer than four satellites are visible.

STRUCTURE

The current GPS consists of three major segments. These are the space segment (SS), a control segment (CS), and a user segment (US). The U.S. Air Force develops, maintains, and operates the space and control segments. GPS satellites broadcast signals from space, and each GPS receiver uses these signals to calculate its three-dimensional location (latitude, longitude, and altitude) and the current time.

The space segment is composed of 24 to 32 satellites in medium Earth orbit and also includes the payload adapters to the boosters required to launch them into orbit. The control segment is composed of a master control station (MCS), an alternate master control station, and a host of dedicated and shared ground antennas and monitor stations. The user segment is composed of hundreds of thousands of U.S. and allied military users of the secure GPS Precise Positioning Service, and hundreds of millions of civil, commercial, and scientific users of the Standard Positioning Service.

Space segment

The space segment (SS) is composed of the orbiting GPS satellites, or Space Vehicles (SV) in GPS parlance. The GPS design originally called for 24 SVs, eight each in three approximately circular orbits, but this was modified to six orbital planes with four satellites each. The six orbit planes have approximately 55° inclination (tilt relative to the Earth's equator) and are separated by 60° right ascension of the ascending node (angle along the equator from a reference point to the orbit's intersection). The orbital period is one-half a sidereal day, i.e., 11 hours and 58 minutes so that the satellites pass over the same locations or almost the same locations every day. The orbits are arranged so that at least six satellites are always within line of sight from almost everywhere on the Earth's surface. The result of this objective is that the four satellites are not evenly spaced (90 degrees) apart within each orbit. In general terms, the angular difference between satellites in each orbit is 30, 105, 120, and 105 degrees apart, which sum to 360 degrees.

Unlaunched GPS block II-A satellite on display at the San Diego Air & Space Museum

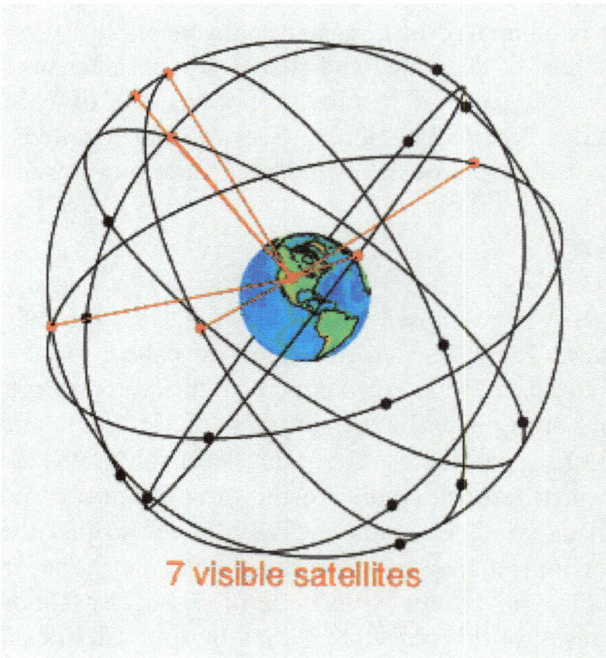

A visual example of a 24 satellite GPS constellation in motion with the earth rotating. Notice how the number of *satellites in view* from a given point on the earth's surface, in this example in Golden CO (39.7469° N, 105.2108° W), changes with time.

Orbiting at an altitude of approximately 20,200 km (12,600 mi); orbital radius of approximately 26,600 km (16,500 mi), each SV makes two complete orbits each sidereal day, repeating the same ground track each day. This was very helpful during development because even with only four satellites, correct alignment means all four are visible from one spot for a few hours each day. For military operations, the ground track repeat can be used to ensure good coverage in combat zones.

As of February 2016, there are 32 satellites in the GPS constellation, 31 of which are in use. The additional satellites improve the precision of GPS receiver calculations by providing redundant measurements. With the increased number of satellites, the constellation was changed to a nonuniform arrangement. Such an arrangement was shown to improve reliability and availability of the system, relative to a uniform system, when multiple satellites fail. About nine satellites are visible from any point on the ground at any one time, ensuring considerable redundancy over the minimum four satellites needed for a position.

Control Segment

The control segment is composed of:

- a master control station (MCS),
- an alternate master control station,
- four dedicated ground antennas, and
- six dedicated monitor stations.

The MCS can also access U.S. Air Force Satellite Control Network (AFSCN) ground antennas (for additional command and control capability) and NGA (National Geospatial-Intelligence Agency) monitor stations. The flight paths of the satellites are tracked by dedicated U.S. Air Force monitoring stations in Hawaii, Kwajalein Atoll, Ascension Island, Diego Garcia, Colorado Springs, Colorado and Cape Canaveral, along with shared NGA monitor stations operated in England, Argentina, Ecuador, Bahrain, Australia and Washington DC. The tracking information is sent to the Air Force Space Command MCS at Schriever Air Force Base 25 km (16 mi) ESE of Colorado Springs, which is operated by the 2nd Space Operations Squadron (2 SOPS) of the U.S. Air Force. Then 2 SOPS contacts each GPS satellite regularly with a navigational update using dedicated or shared (AFSCN) ground antennas (GPS dedicated ground antennas are located at Kwajalein, Ascension Island, Diego Garcia, and Cape Canaveral). These updates synchronize the atomic clocks on board the satellites to within a few nanoseconds of each other, and adjust the ephemeris of each satellite's internal orbital model. The updates are created by a Kalman filter that uses inputs from the ground monitoring stations, space weather information, and various other inputs.

Ground monitor station used from 1984 to 2007, on display at the Air Force Space & Missile Museum.

Satellite maneuvers are not precise by GPS standards—so to change a satellite's orbit, the satellite must be marked *unhealthy*, so receivers don't use it. After the satellite maneuver, engineers track the new orbit from the ground, upload the new ephemeris, and mark the satellite healthy again.

The Operation Control Segment (OCS) currently serves as the control segment of record. It provides the operational capability that supports GPS users and keeps the GPS system operational and performing within specification.

OCS successfully replaced the legacy 1970s-era mainframe computer at Schriever Air Force Base in September 2007. After installation, the system helped enable upgrades and provide a foundation for a new security architecture that supported U.S. armed forces. OCS will continue to be the ground control system of record until the new segment, Next Generation GPS Operation Control System (OCX), is fully developed and functional.

The new capabilities provided by OCX will be the cornerstone for revolutionizing GPS's mission capabilities, enabling Air Force Space Command to greatly enhance GPS operational services to U.S. combat forces, civil partners and myriad domestic and international users.

The GPS OCX program also will reduce cost, schedule and technical risk. It is designed to provide 50% sustainment cost savings through efficient software architecture and Performance-Based Logistics. In addition, GPS OCX is expected to cost millions less than the cost to upgrade OCS while providing four times the capability.

The GPS OCX program represents a critical part of GPS modernization and provides significant information assurance improvements over the current GPS OCS program.

- OCX will have the ability to control and manage GPS legacy satellites as well as the next generation of GPS III satellites, while enabling the full array of military signals.

- Built on a flexible architecture that can rapidly adapt to the changing needs of today's and future GPS users allowing immediate access to GPS data and constellation status through secure, accurate and reliable information.
- Provides the warfighter with more secure, actionable and predictive information to enhance situational awareness.
- Enables new modernized signals (L1C, L2C, and L5) and has M-code capability, which the legacy system is unable to do.
- Provides significant information assurance improvements over the current program including detecting and preventing cyber attacks, while isolating, containing and operating during such attacks.
- Supports higher volume near real-time command and control capabilities and abilities.

On September 14, 2011, the U.S. Air Force announced the completion of GPS OCX Preliminary Design Review and confirmed that the OCX program is ready for the next phase of development.

The GPS OCX program has missed major milestones and is pushing the GPS IIIA launch beyond April 2016.

User Segment

The user segment is composed of hundreds of thousands of U.S. and allied military users of the secure GPS Precise Positioning Service, and tens of millions of civil, commercial and scientific users of the Standard Positioning Service. In general, GPS receivers are composed of an antenna, tuned to the frequencies transmitted by the satellites, receiver-processors, and a highly stable clock (often a crystal oscillator). They may also include a display for providing location and speed information to the user.

GPS receivers come in a variety of formats, from devices integrated into cars, phones, and watches, to dedicated devices such as these.

The first portable GPS unit, Leica WM 101 displayed at the Irish National Science Museum at Maynooth.

A receiver is often described by its number of channels: this signifies how many satellites it can monitor simultaneously. Originally limited to four or five, this has progressively increased over the years so that, as of 2007, receivers typically have between 12 and 20 channels.

A typical OEM GPS receiver module measuring 15×17 mm.

GPS receivers may include an input for differential corrections, using the RTCM SC-104 format. This is typically in the form of an RS-232 port at 4,800 bit/s speed. Data is actually sent at a much lower rate, which limits the accuracy of the signal sent using

RTCM. Receivers with internal DGPS receivers can outperform those using external RTCM data. As of 2006, even low-cost units commonly include Wide Area Augmentation System (WAAS) receivers.

A typical GPS receiver with integrated antenna.

Many GPS receivers can relay position data to a PC or other device using the NMEA 0183 protocol. Although this protocol is officially defined by the National Marine Electronics Association (NMEA), references to this protocol have been compiled from public records, allowing open source tools like gpsd to read the protocol without violating intellectual property laws. Other proprietary protocols exist as well, such as the SiRF and MTK protocols. Receivers can interface with other devices using methods including a serial connection, USB, or Bluetooth.

APPLICATIONS

While originally a military project, GPS is considered a *dual-use* technology, meaning it has significant military and civilian applications.

GPS has become a widely deployed and useful tool for commerce, scientific uses, tracking, and surveillance. GPS's accurate time facilitates everyday activities such as banking, mobile phone operations, and even the control of power grids by allowing well synchronized hand-off switching.

Civilian

Many civilian applications use one or more of GPS's three basic components: absolute location, relative movement, and time transfer.

- Agriculture:GPS has made a great evolution in different aspects of modern agricultural sectors. Today, a growing number of crop producers are using GPS and other modern electronic and computer equipment to practice Site Specific Management (SSM) and precision agriculture. This technology has the potential in agricultural mechanization (farm and machinery management) by providing farmers with a sophisticated tool to measure yield on much smaller scales as well as precisely determination and automatic storing of variables such as field time, working area, machine travel distance and speed, fuel consumption and yield information.
- Astronomy: both positional and clock synchronization data is used in astrometry and celestial mechanics. GPS is also used in both amateur astronomy with small telescopes as well as by professional observatories for finding extrasolar planets, for example.
- Automated vehicle: applying location and routes for cars and trucks to function without a human driver.
- Cartography: both civilian and military cartographers use GPS extensively.

This antenna is mounted on the roof of a hut containing a scientific experiment needing precise timing.

- Cellular telephony: clock synchronization enables time transfer, which is critical for synchronizing its spreading codes with other base stations to facilitate inter-cell handoff and support hybrid GPS/cellular position detection for mobile emergency calls and other applications. The first handsets with integrated GPS launched in the late 1990s. The U.S. Federal Communications Commission (FCC) mandated the feature in either the handset or in the towers (for use in triangulation) in 2002 so emergency services could locate 911 callers. Third-party software developers later gained access to GPS APIs from Nextel upon launch, followed by Sprint in 2006, and Verizon soon thereafter.
- Clock synchronization: the accuracy of GPS time signals (±10 ns) is second only to the atomic clocks they are based on.
- Disaster relief/emergency services: many emergency services depend upon GPS for location and timing capabilities.
- GPS-equipped radiosondes and dropsondes: measure and calculate the atmospheric pressure, wind speed and direction up to 27 km from the Earth's surface.
- Radio occultation for weather and atmospheric science applications.

- Fleet tracking: used to identify, locate and maintain contact reports with one or more fleet vehicles in real-time.
- Geofencing: vehicle tracking systems, person tracking systems, and pet tracking systems use GPS to locate devices that are attached to or carried by a person, vehicle, or pet. The application can provide continuous tracking and send notifications if the target leaves a designated (or “fenced-in”) area.
- Geotagging: applies location coordinates to digital objects such as photographs (in Exif data) and other documents for purposes such as creating map overlays with devices like Nikon GP-1
- GPS aircraft tracking
- GPS for mining: the use of RTK GPS has significantly improved several mining operations such as drilling, shoveling, vehicle tracking, and surveying. RTK GPS provides centimeter-level positioning accuracy.
- GPS data mining: It is possible to aggregate GPS data from multiple users to understand movement patterns, common trajectories and interesting locations.
- GPS tours: location determines what content to display; for instance, information about an approaching point of interest.
- Navigation: navigators value digitally precise velocity and orientation measurements.
- Phasor measurements: GPS enables highly accurate timestamping of power system measurements, making it possible to compute phasors.
- Recreation: for example, Geocaching, Geodashing, GPS drawing, waymarking, and other kinds of location based mobile games.
- Robotics: self-navigating, autonomous robots using a GPS sensors, which calculate latitude, longitude, time, speed, and heading.
- Sport: used in football and rugby for the control and analysis of the training load.
- Surveying: surveyors use absolute locations to make maps and determine property boundaries.
- Tectonics: GPS enables direct fault motion measurement of earthquakes. Between earthquakes GPS can be used to measure crustal motion and deformation to estimate seismic strain buildup for creating seismic hazard maps.
- Telematics: GPS technology integrated with computers and mobile communications technology in automotive navigation systems.

Restrictions on Civilian Use

The U.S. government controls the export of some civilian receivers. All GPS receivers capable of functioning above 18 km (60,000 feet) altitude and 515 m/s (1,000 knots), or designed or modified for use with unmanned air vehicles like, e.g., ballistic or cruise missile systems, are classified as munitions (weapons)—which means they require State Department export licenses.

This rule applies even to otherwise purely civilian units that only receive the L1 frequency and the C/A (Coarse/Acquisition) code.

Disabling operation above these limits exempts the receiver from classification as a munition. Vendor interpretations differ. The rule refers to operation at both the target altitude and speed, but some receivers stop operating even when stationary. This has caused problems with some amateur radio balloon launches that regularly reach 30 km (100,000 feet).

These limits only apply to units or components exported from the USA. A growing trade in various components exists, including GPS units from other countries. These are expressly sold as ITAR-free.

Military

Attaching a GPS guidance kit to a dumb bomb, March 2003.

M982 Excalibur GPS-guided artillery shell.

As of 2009, military GPS applications include:

- Navigation: Soldiers use GPS to find objectives, even in the dark or in unfamiliar territory, and to coordinate troop and supply movement. In the United States armed forces, commanders use the *Commanders Digital Assistant* and lower ranks use the *Soldier Digital Assistant*.
- Target tracking: Various military weapons systems use GPS to track potential ground and air targets before flagging them as hostile. These weapon systems

pass target coordinates to precision-guided munitions to allow them to engage targets accurately. Military aircraft, particularly in air-to-ground roles, use GPS to find targets.

- Missile and projectile guidance: GPS allows accurate targeting of various military weapons including ICBMs, cruise missiles, precision-guided munitions and Artillery projectiles. Embedded GPS receivers able to withstand accelerations of 12,000 g or about 118 km/s^2 have been developed for use in 155-millimeter (6.1 in) howitzer shells.
- Search and rescue.
- Reconnaissance: Patrol movement can be managed more closely.
- GPS satellites carry a set of nuclear detonation detectors consisting of an optical sensor (Y-sensor), an X-ray sensor, a dosimeter, and an electromagnetic pulse (EMP) sensor (W-sensor), that form a major portion of the United States Nuclear Detonation Detection System. General William Shelton has stated that future satellites may drop this feature to save money.

GPS type navigation was first used in war in the 1991 Persian Gulf War, before GPS was fully developed in 1995, to assist Coalition Forces to navigate and perform maneuvers in the war. The war also demonstrated the vulnerability of GPS to being jammed, when Iraqi forces added noise to the weak GPS signal transmission to protect Iraqi targets.

Communication

The navigational signals transmitted by GPS satellites encode a variety of information including satellite positions, the state of the internal clocks, and the health of the network. These signals are transmitted on two separate carrier frequencies that are common to all satellites in the network. Two different encodings are used: a public encoding that enables lower resolution navigation, and an encrypted encoding used by the U.S. military.

Message Format

GPS message format	
Subframes	**Description**
1	Satellite clock, GPS time relationship
2–3	Ephemeris (precise satellite orbit)
4–5	Almanac component (satellite network synopsis, error correction)

Each GPS satellite continuously broadcasts a *navigation message* on L1 (C/A and P/Y) and L2 (P/Y) frequencies at a rate of 50 bits per second. Each complete message takes 750 seconds (12 1/2 minutes) to complete. The message structure has a basic format of a 1500-bit-long frame made up of five subframes, each subframe being 300 bits (6 seconds) long. Subframes 4 and 5 are subcommutated 25 times each, so that a complete data message requires the transmission of 25 full frames. Each subframe consists of ten words, each 30 bits long. Thus, with 300 bits in a subframe times 5 subframes in a frame times 25 frames in a message, each message is 37,500 bits long. At a transmission rate of 50-bit/s, this gives 750 seconds to transmit an entire almanac message (GPS). Each 30-second frame begins precisely on the minute or half-minute as indicated by the atomic clock on each satellite.

The first subframe of each frame encodes the week number and the time within the week, as well as the data about the health of the satellite. The second and the third subframes contain the *ephemeris* – the precise orbit for the satellite. The fourth and fifth subframes contain the *almanac*, which contains coarse orbit and status information for up to 32 satellites in the constellation as well as data related to error correction. Thus, to obtain an accurate satellite location from this transmitted message, the receiver must demodulate the message from each satellite it includes in its solution for 18 to 30 seconds. To collect all transmitted almanacs, the receiver must demodulate the message for 732 to 750 seconds or 12 1/2 minutes.

All satellites broadcast at the same frequencies, encoding signals using unique code division multiple access (CDMA) so receivers can distinguish individual satellites from each other. The system uses two distinct CDMA encoding types: the coarse/acquisition (C/A) code, which is accessible by the general public, and the precise (P(Y)) code, which is encrypted so that only the U.S. military and other NATO nations who have been given access to the encryption code can access it.

The ephemeris is updated every 2 hours and is generally valid for 4 hours, with provisions for updates every 6 hours or longer in non-nominal conditions. The almanac is updated typically every 24 hours. Additionally, data for a few weeks following is uploaded in case of transmission updates that delay data upload.

Satellite Frequencies

All satellites broadcast at the same two frequencies, 1.57542 GHz (L1 signal) and 1.2276 GHz (L2 signal). The satellite network uses a CDMA spread-spectrum technique where the low-bitrate message data is encoded with a high-rate pseudo-random (PRN) sequence that is different for each satellite. The receiver must be aware of the PRN codes for each satellite to reconstruct the actual message data. The C/A code, for civilian use, transmits data at 1.023 million chips per second, whereas the P code, for U.S. military use, transmits at 10.23 million chips per second. The actual internal reference of the satellites is 10.22999999543 MHz to compensate for relativistic effects that make observers on the Earth perceive a different time reference with respect to the transmitters in orbit. The L1 carrier is modulated by both the C/A and P codes, while the L2 carrier is only modulated by the P code. The P code can be encrypted as a so-called P(Y) code that is only available to military equipment with a proper decryption key. Both the C/A and P(Y) codes impart the precise time-of-day to the user.

GPS frequency overview		
Band	**Frequency**	**Description**
L1	1575.42 MHz	Coarse-acquisition (C/A) and encrypted precision (P(Y)) codes, plus the L1 civilian (L1C) and military (M) codes on future Block III satellites.
L2	1227.60 MHz	P(Y) code, plus the L2C and military codes on the Block IIR-M and newer satellites.
L3	1381.05 MHz	Used for nuclear detonation (NUDET) detection.
L4	1379.913 MHz	Being studied for additional ionospheric correction.
L5	1176.45 MHz	Proposed for use as a civilian safety-of-life (SoL) signal.

The L3 signal at a frequency of 1.38105 GHz is used to transmit data from the satellites to ground stations. This data is used by the United States Nuclear Detonation (NUDET) Detection System (USNDS) to detect, locate, and report nuclear detonations (NUDETs) in the Earth's atmosphere and near space. One usage is the enforcement of nuclear test ban treaties.

The L4 band at 1.379913 GHz is being studied for additional ionospheric correction.

The L5 frequency band at 1.17645 GHz was added in the process of GPS modernization. This frequency falls into an internationally protected range for aeronautical navigation, promising little or no interference under all circumstances. The first Block IIF satellite that provides this signal was launched in 2010. The L5 consists of two carrier components that are in phase quadrature with each other. Each carrier component is bi-phase shift key (BPSK) modulated by a separate bit train. "L5, the third civil GPS signal, will eventually support safety-of-life applications for aviation and provide improved availability and accuracy."

A conditional waiver has recently (2011-01-26) been granted to LightSquared to operate a terrestrial broadband service near the L1 band. Although LightSquared had applied for a license to operate in the 1525 to 1559 band as early as 2003 and it was put out for public comment, the FCC asked LightSquared to form a study group with the GPS community to test GPS receivers and identify issue that might arise due to the larger signal power from the LightSquared terrestrial network. The GPS community had not objected to the LightSquared (formerly MSV and SkyTerra) applications until November 2010, when LightSquared applied for a modification to its Ancillary Terrestrial Component (ATC) authorization. This filing (SAT-MOD-20101118-00239) amounted to a request to run several orders of magnitude more power in the same frequency band for terrestrial base stations, essentially repurposing what was supposed to be a "quiet neighborhood" for signals from space as the equivalent of a cellular network. Testing in the first half of 2011 has demonstrated that the impact of the lower 10 MHz of spectrum is minimal to GPS devices (less than 1% of the total GPS devices are affected). The upper 10 MHz intended for use by LightSquared may have some impact on GPS devices. There is some concern that this may seriously degrade the GPS signal for many consumer uses. Aviation Week magazine reports that the latest testing (June 2011) confirms "significant jamming" of GPS by LightSquared's system.

Demodulation and Decoding

Because all of the satellite signals are modulated onto the same L1 carrier frequency, the signals must be separated after demodulation. This is done by assigning each satellite a unique binary sequence known as a Gold code. The signals are decoded after demodulation using addition of the Gold codes corresponding to the satellites monitored by the receiver.

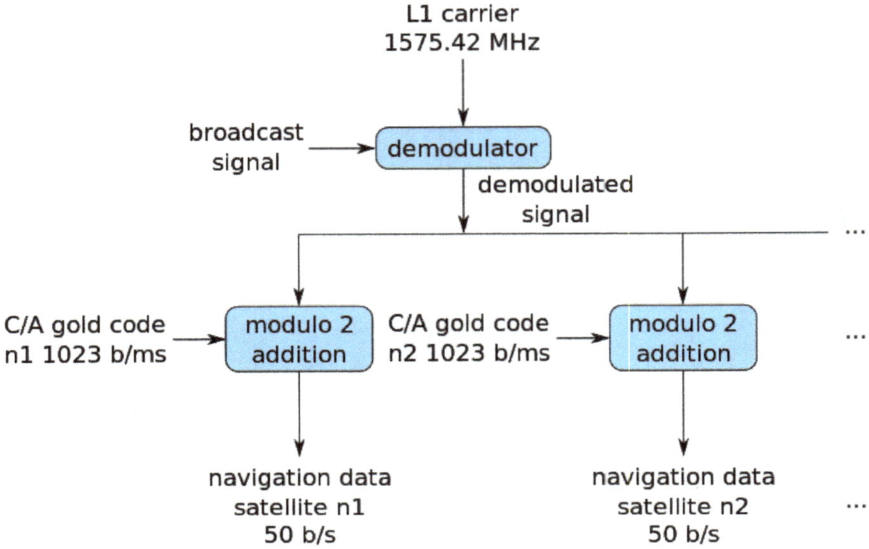

Demodulating and Decoding GPS Satellite Signals using the Coarse/Acquisition Gold code.

If the almanac information has previously been acquired, the receiver picks the satellites to listen for by their PRNs, unique numbers in the range 1 through 32. If the almanac information is not in memory, the receiver enters a search mode until a lock is obtained on one of the satellites. To obtain a lock, it is necessary that there be an unobstructed line of sight from the receiver to the satellite. The receiver can then acquire the almanac and determine the satellites it should listen for. As it detects each satellite's signal, it identifies it by its distinct C/A code pattern. There can be a delay of up to 30 seconds before the first estimate of position because of the need to read the ephemeris data.

Processing of the navigation message enables the determination of the time of transmission and the satellite position at this time.

Navigation Equations

Problem Description

The receiver uses messages received from satellites to determine the satellite positions and time sent. The *x, y,* and *z* components of satellite position and the time sent are designated as $[x_i, y_i, z_i, s_i]$ where the subscript *i* denotes the satellite and has the value

1, 2, ..., n, where $n \geq 4$. When the time of message reception indicated by the on-board receiver clock is $\tilde{t}_i$, the true reception time is $t_i = \tilde{t}_i - b$, where b is the receiver's clock bias from the much more accurate GPS system clocks employed by the satellites. The receiver clock bias is the same for all received satellite signals (assuming the satellite clocks are all perfectly synchronized). The message's transit time is $\tilde{t}_i - b - s_i$, where s_i is the satellite time. Assuming the message traveled at the speed of light, c, the distance traveled is $(\tilde{t}_i - b - s_i)\ c$.

For n satellites, the equations to satisfy are:

$$(x - x_i)^2 + (y - y_i)^2 + (z - z_i)^2 = \left([\tilde{t}_i - b - s_i]c\right)^2,\ i = 1, 2, \ldots, n$$

or in terms of *pseudoranges*, $p_i = \left(\tilde{t}_i - s_i\right) c$, as

$$\sqrt{(x - x_i)^2 + (y - y_i)^2 + (z - z_i)^2} + bc = p_i,\ i = 1, 2, \ldots, n$$

Since the equations have four unknowns [x, y, z, b]—the three components of GPS receiver position and the clock bias—signals from at least four satellites are necessary to attempt solving these equations. They can be solved by algebraic or numerical methods. Existence and uniqueness of GPS solutions are discussed by Abell and Chaffee. When n is greater than 4 this system is overdetermined and a fitting method must be used.

With each combination of satellites, GDOP quantities can be calculated based on the relative sky directions of the satellites used. The receiver location is expressed in a specific coordinate system, such as latitude and longitude using the WGS 84 geodetic datum or a country-specific system.

Geometric Interpretation

The GPS equations can be solved by numerical and analytical methods. Geometrical interpretations can enhance the understanding of these solution methods.

Spheres

The measured ranges, called pseudoranges, contain clock errors. In a simplified idealization in which the ranges are synchronized, these true ranges represent the radii of spheres, each centered on one of the transmitting satellites. The solution for the position of the receiver is then at the intersection of the surfaces of three of these spheres. If more than the minimum number of ranges is available, a near intersection of more than three sphere surfaces could be found via, e.g. least squares.

Hyperboloids

If the distance traveled between the receiver and satellite i and the distance traveled between the receiver and satellite j are subtracted, the result is $(\tilde{t}_i - s_i)\ c - (\tilde{t}_j - s_j)\ c$, which only involves known or measured quantities. The locus of points having a constant difference in distance to two points (here, two satellites) is a hyperboloid. Thus, from four or more measured reception times, the receiver can be placed at the intersection of the surfaces of three or more hyperboloids.

Spherical Cones

The solution space [x, y, z, b] can be seen as a four-dimensional geometric space. In that case each of the equations describes a spherical cone, with the cusp located at the satellite, and the base a sphere around the satellite. The receiver is at the intersection of four or more of such cones.

Solution Methods

Least Squares

When more than four satellites are available, the calculation can use the four best, or more than four simultaneously (up to all visible satellites), depending on the number of receiver channels, processing capability, and geometric dilution of precision (GDOP).

Using more than four involves an over-determined system of equations with no unique solution; such a system can be solved by a least-squares or weighted least squares method.

$$\left(\hat{x}, \hat{y}, \hat{z}, \hat{b}\right) = \underset{(x,y,z,b)}{\arg\min} \sum_i \left(\sqrt{(x - x_i)^2 + (y - y_i)^2 + (z - z_i)^2} + bc - p_i\right)^2$$

Iterative

Both the equations for four satellites, or the least squares equations for more than four, are non-linear and need special solution methods. A common approach is by iteration on a linearized form of the equations, such as the Gauss–Newton algorithm.

The GPS system was initially developed assuming use of a numerical least-squares solution method—i.e., before closed-form solutions were found.

Closed-Form

One closed-form solution to the above set of equations was developed by S. Bancroft. Its properties are well known; in particular, proponents claim it is superior in low-GDOP situations, compared to iterative least squares methods.

Bancroft's method is algebraic, as opposed to numerical, and can be used for four or more satellites. When four satellites are used, the key steps are inversion of a 4x4 matrix and solution of a single-variable quadratic equation. Bancroft's method provides one or two solutions for the unknown quantities. When there are two (usually the case), only one is a near-Earth sensible solution.

When a receiver uses more than four satellites for a solution, Bancroft uses the generalized inverse (i.e., the pseudoinverse) to find a solution. However, a case has been made that iterative methods (e.g., Gauss–Newton algorithm) for solving over-determined non-linear least squares (NLLS) problems generally provide more accurate solutions.

Leick et al. (2015) states that "Bancroft's (1985) solution is a very early, if not the first, closed-form solution." Other closed-form solutions were published afterwards, although their adoption in practice is unclear.

Error Sources and Analysis

GPS error analysis examines error sources in GPS results and the expected size of those errors. GPS makes corrections for receiver clock errors and other effects, but some residual errors remain uncorrected. Error sources include signal arrival time measurements, numerical calculations, atmospheric effects (ionospheric/tropospheric delays), ephemeris and clock data, multipath signals, and natural and artificial interference. Magnitude of residual errors from these sources depends on geometric dilution of precision. Artificial errors may result from jamming devices and threaten ships and aircraft or from intentional signal degradation through selective availability, which limited accuracy to ~6–12 m, but has been switched off since May 1, 2000.

Accuracy Enhancement and Surveying

Augmentation

Integrating external information into the calculation process can materially improve accuracy. Such augmentation systems are generally named or described based on how the information arrives. Some systems transmit additional error information (such as clock drift, ephemera, or ionospheric delay), others characterize prior errors, while a third group provides additional navigational or vehicle information.

Examples of augmentation systems include the Wide Area Augmentation System (WAAS), European Geostationary Navigation Overlay Service (EGNOS), Differential GPS (DGPS), inertial navigation systems (INS) and Assisted GPS. The standard accuracy of about 15 meters (49 feet) can be augmented to 3–5 meters (9.8–16.4 ft) with DGPS, and to about 3 meters (9.8 feet) with WAAS.

Precise Monitoring

Accuracy can be improved through precise monitoring and measurement of existing GPS signals in additional or alternate ways.

The largest remaining error is usually the unpredictable delay through the ionosphere. The spacecraft broadcast ionospheric model parameters, but some errors remain. This is one reason GPS spacecraft transmit on at least two frequencies, L1 and L2. Ionospheric delay is a well-defined function of frequency and the total electron content (TEC) along the path, so measuring the arrival time difference between the frequencies determines TEC and thus the precise ionospheric delay at each frequency.

Military receivers can decode the P(Y) code transmitted on both L1 and L2. Without decryption keys, it is still possible to use a *codeless* technique to compare the P(Y) codes on L1 and L2 to gain much of the same error information. However, this technique is slow, so it is currently available only on specialized surveying equipment. In the future, additional civilian codes are expected to be transmitted on the L2 and L5 frequencies. All users will then be able to perform dual-frequency measurements and directly compute ionospheric delay errors.

A second form of precise monitoring is called *Carrier-Phase Enhancement* (CPGPS). This corrects the error that arises because the pulse transition of the PRN is not instantaneous, and thus the correlation (satellite–receiver sequence matching) operation is imperfect. CPGPS uses the L1 carrier wave, which has a period of $\frac{1\,\mathrm{s}}{1575.42 \times 10^6} = 0.63475\,\mathrm{ns} \approx 1\,\mathrm{ns}$, which is about one-thousandth of the C/A Gold code bit period of $\frac{1\,\mathrm{s}}{1023 \times 10^3} = 977.5\,\mathrm{ns} \approx 1000\,\mathrm{ns}$, to act as an additional clock signal and resolve the uncertainty. The phase difference error in the normal GPS amounts to 2–3 meters (7–10 ft) of ambiguity. CPGPS working to within 1% of perfect transition reduces this error to 3 centimeters (1.2 in) of ambiguity. By eliminating this error source, CPGPS coupled with DGPS normally realizes between 20–30 centimeters (8–12 in) of absolute accuracy.

Relative Kinematic Positioning (RKP) is a third alternative for a precise GPS-based positioning system. In this approach, determination of range signal can be resolved to a precision of less than 10 centimeters (4 in). This is done by resolving the number of cycles that the signal is transmitted and received by the receiver by using a combination of differential GPS (DGPS) correction data, transmitting GPS signal phase information and ambiguity resolution techniques via statistical tests—possibly with processing in real-time (real-time kinematic positioning, RTK).

Timekeeping

Leap Seconds

While most clocks derive their time from Coordinated Universal Time (UTC), the atomic clocks on the satellites are set to GPS time (GPST). The difference is that GPS time is not corrected to match the rotation of the Earth, so it does not contain leap seconds or other corrections that are periodically added to UTC. GPS time was set to match UTC in 1980, but has since diverged. The lack of corrections means that GPS time remains at a constant offset with International Atomic Time (TAI) (TAI − GPS = 19 seconds). Periodic corrections are performed to the on-board clocks to keep them synchronized with ground clocks.

The GPS navigation message includes the difference between GPS time and UTC. As of July 2015, GPS time is 17 seconds ahead of UTC because of the leap second added to UTC on June 30, 2015. Receivers subtract this offset from GPS time to calculate UTC and specific timezone values. New GPS units may not show the correct UTC time until after receiving the UTC offset message. The GPS-UTC offset field can accommodate 255 leap seconds (eight bits).

Accuracy

GPS time is theoretically accurate to about 14 nanoseconds. However, most receivers lose accuracy in the interpretation of the signals and are only accurate to 100 nanoseconds.

Format

As opposed to the year, month, and day format of the Gregorian calendar, the GPS date is expressed as a week number and a seconds-into-week number. The week number is transmitted as a ten-bit field in the C/A and P(Y) navigation messages, and so it becomes zero again every 1,024 weeks (19.6 years). GPS week zero started at 00:00:00 UTC (00:00:19 TAI) on January 6, 1980, and the week number became zero again for the first time at 23:59:47 UTC on August 21, 1999 (00:00:19 TAI on August 22, 1999). To determine the current Gregorian date, a GPS receiver must be provided with the approximate date (to within 3,584 days) to correctly translate the GPS date signal. To address this concern the modernized GPS navigation message uses a 13-bit field that only repeats every 8,192 weeks (157 years), thus lasting until the year 2137 (157 years after GPS week zero).

Carrier Phase Tracking (Surveying)

Another method that is used in surveying applications is carrier phase tracking. The period of the carrier frequency multiplied by the speed of light gives the wavelength, which is about 0.19 meters for the L1 carrier. Accuracy within 1% of wavelength in detecting the leading edge reduces this component of pseudorange error to as little as 2 millimeters. This compares to 3 meters for the C/A code and 0.3 meters for the P code.

However, 2 millimeter accuracy requires measuring the total phase—the number of waves multiplied by the wavelength plus the fractional wavelength, which requires specially equipped receivers. This method has many surveying applications. It is accurate enough for real-time tracking of the very slow motions of tectonic plates, typically 0–100 mm (0–4 inches) per year.

Triple differencing followed by numerical root finding, and a mathematical technique called least squares can estimate the position of one receiver given the position of another. First, compute the difference between satellites, then between receivers, and finally between epochs. Other orders of taking differences are equally valid. Detailed discussion of the errors is omitted.

The satellite carrier total phase can be measured with ambiguity as to the number of cycles. Let $\phi(r_i, s_j, t_k)$ denote the phase of the carrier of satellite j measured by receiver i at time t_k. This notation shows the meaning of the subscripts i, j, and k. The receiver (r), satellite (s), and time (t) come in alphabetical order as arguments of ϕ and to balance readability and conciseness, let $\phi_{i,j,k} = \phi(r_i, s_j, t_k)$ be a concise abbreviation. Also we define three functions, $\Delta^r, \Delta^s, \Delta^t$,:, which return differences between receivers, satellites, and time points, respectively. Each function has variables with three subscripts as its arguments. These three functions are defined below. If $\alpha_{i,j,k}$ is a function of the three integer arguments, i, j, and k then it is a valid argument for the functions,: $\Delta^r, \Delta^s, \Delta^t$ with the values defined as

$$\Delta^r(\alpha_{i,j,k}) = \alpha_{i+1,j,k} - \alpha_{i,j,k},$$
$$\Delta^s(\alpha_{i,j,k}) = \alpha_{i,j+1,k} - \alpha_{i,j,k}, \text{ and}$$
$$\Delta^t(\alpha_{i,j,k}) = \alpha_{i,j,k+1} - \alpha_{i,j,k}.$$

Also if $\alpha_{i,j,k}$ *and* $\beta_{l,m,n}$ are valid arguments for the three functions and a and b are constants then $(a\,\alpha_{i,j,k} + b\,\beta_{l,m,n})$ is a valid argument with values defined as

$$\Delta^r(a\,\alpha_{i,j,k} + b\,\beta_{l,m,n}) = a\,\Delta^r(\alpha_{i,j,k}) + b\,\Delta^r(\beta_{l,m,n}),$$
$$\Delta^s(a\,\alpha_{i,j,k} + b\,\beta_{l,m,n}) = a\,\Delta^s(\alpha_{i,j,k}) + b\,\Delta^s(\beta_{l,m,n}), \text{ and}$$
$$\Delta^t(a\,\alpha_{i,j,k} + b\,\beta_{l,m,n}) = a\,\Delta^t(\alpha_{i,j,k}) + b\,\Delta^t(\beta_{l,m,n}).$$

Receiver clock errors can be approximately eliminated by differencing the phases measured from satellite 1 with that from satellite 2 at the same epoch. This difference is designated as $\Delta^s(\phi_{1,1,1}) = \phi_{1,2,1} - \phi_{1,1,1}$

Double differencing computes the difference of receiver 1's satellite difference from that of receiver 2. This approximately eliminates satellite clock errors. This double difference is:

$$\Delta^r(\Delta^s(\phi_{1,1,1})) = \Delta^r(\phi_{1,2,1} - \phi_{1,1,1}) \qquad = \Delta^r(\phi_{1,2,1}) - \Delta^r(\phi_{1,1,1}) = (\phi_{2,2,1} - \phi_{1,2,1}) - (\phi_{2,1,1} - \phi_{1,1,1})$$

Triple differencing subtracts the receiver difference from time 1 from that of time 2. This eliminates the ambiguity associated with the integral number of wavelengths in carrier phase provided this ambiguity does not change with time. Thus the triple difference result eliminates practically all clock bias errors and the integer ambiguity. Atmospheric delay and satellite ephemeris errors have been significantly reduced. This triple difference is:

$$\Delta^t(\Delta^r(\Delta^s(\phi_{1,1,1})))$$

Triple difference results can be used to estimate unknown variables. For example, if the position of receiver 1 is known but the position of receiver 2 unknown, it may be possible to estimate the position of receiver 2 using numerical root finding and least squares. Triple difference results for three independent time pairs may be sufficient to solve for receiver 2's three position components. This may require a numerical procedure. An approximation of receiver 2's position is required to use such a numerical method. This initial value can probably be provided from the navigation message and the intersection of sphere surfaces. Such a reasonable estimate can be key to successful multidimensional root finding. Iterating from three time pairs and a fairly good initial value produces one observed triple difference result for receiver 2's position. Processing additional time pairs can improve accuracy, overdetermining the answer with multiple solutions. Least squares can estimate an overdetermined system. Least squares determines the position of receiver 2 that best fits the observed triple difference results for receiver 2 positions under the criterion of minimizing the sum of the squares.

Regulatory Spectrum Issues Concerning GPS Receivers

In the United States, GPS receivers are regulated under the Federal Communications

Commission's (FCC) Part 15 rules. As indicated in the manuals of GPS-enabled devices sold in the United States, as a Part 15 device, it "must accept any interference received, including interference that may cause undesired operation." With respect to GPS devices in particular, the FCC states that GPS receiver manufacturers, "must use receivers that reasonably discriminate against reception of signals outside their allocated spectrum." For the last 30 years, GPS receivers have operated next to the Mobile Satellite Service band, and have discriminated against reception of mobile satellite services, such as Inmarsat, without any issue.

The spectrum allocated for GPS L1 use by the FCC is 1559 to 1610 MHz, while the spectrum allocated for satellite-to-ground use owned by LightSquared is the Mobile Satellite Service band. Since 1996, the FCC has authorized licensed use of the spectrum neighboring the GPS band of 1525 to 1559 MHz to the Virginia company LightSquared. On March 1, 2001, the FCC received an application from LightSquared's predecessor, Motient Services to use their allocated frequencies for an integrated satellite-terrestrial service. In 2002, the U.S. GPS Industry Council came to an out-of-band-emissions (OOBE) agreement with LightSquared to prevent transmissions from LightSquared's ground-based stations from emitting transmissions into the neighboring GPS band of 1559 to 1610 MHz. In 2004, the FCC adopted the OOBE agreement in its authorization for LightSquared to deploy a ground-based network ancillary to their satellite system – known as the Ancillary Tower Components (ATCs) – "We will authorize MSS ATC subject to conditions that ensure that the added terrestrial component remains ancillary to the principal MSS offering. We do not intend, nor will we permit, the terrestrial component to become a stand-alone service." This authorization was reviewed and approved by the U.S. Interdepartment Radio Advisory Committee, which includes the U.S. Department of Agriculture, U.S. Air Force, U.S. Army, U.S. Coast Guard, Federal Aviation Administration, National Aeronautics and Space Administration, Interior, and U.S. Department of Transportation.

In January 2011, the FCC conditionally authorized LightSquared's wholesale customers—such as Best Buy, Sharp, and C Spire—to only purchase an integrated satellite-ground-based service from LightSquared and re-sell that integrated service on devices that are equipped to only use the ground-based signal using LightSquared's allocated frequencies of 1525 to 1559 MHz. In December 2010, GPS receiver manufacturers expressed concerns to the FCC that LightSquared's signal would interfere with GPS receiver devices although the FCC's policy considerations leading up to the January 2011 order did not pertain to any proposed changes to the maximum number of ground-based LightSquared stations or the maximum power at which these stations could operate. The January 2011 order makes final authorization contingent upon studies of GPS interference issues carried out by a LightSquared led working group along with GPS industry and Federal agency participation. On February 14, 2012, the FCC initiated proceedings to vacate LightSquared's Conditional Waiver Order based on the NTIA's conclusion that there was currently no practical way to mitigate potential GPS interference.

GPS receiver manufacturers design GPS receivers to use spectrum beyond the GPS-allocated band. In some cases, GPS receivers are designed to use up to 400 MHz of spectrum in either direction of the L1 frequency of 1575.42 MHz, because mobile satellite services in those regions are broadcasting from space to ground, and at power levels commensurate with mobile satellite services. However, as regulated under the FCC's Part

15 rules, GPS receivers are not warranted protection from signals outside GPS-allocated spectrum. This is why GPS operates next to the Mobile Satellite Service band, and also why the Mobile Satellite Service band operates next to GPS. The symbiotic relationship of spectrum allocation ensures that users of both bands are able to operate cooperatively and freely.

The FCC adopted rules in February 2003 that allowed Mobile Satellite Service (MSS) licensees such as LightSquared to construct a small number of ancillary ground-based towers in their licensed spectrum to "promote more efficient use of terrestrial wireless spectrum." In those 2003 rules, the FCC stated "As a preliminary matter, terrestrial [Commercial Mobile Radio Service ("CMRS")] and MSS ATC are expected to have different prices, coverage, product acceptance and distribution; therefore, the two services appear, at best, to be imperfect substitutes for one another that would be operating in predominately different market segments... MSS ATC is unlikely to compete directly with terrestrial CMRS for the same customer base...". In 2004, the FCC clarified that the ground-based towers would be ancillary, noting that "We will authorize MSS ATC subject to conditions that ensure that the added terrestrial component remains ancillary to the principal MSS offering. We do not intend, nor will we permit, the terrestrial component to become a stand-alone service." In July 2010, the FCC stated that it expected LightSquared to use its authority to offer an integrated satellite-terrestrial service to "provide mobile broadband services similar to those provided by terrestrial mobile providers and enhance competition in the mobile broadband sector." However, GPS receiver manufacturers have argued that LightSquared's licensed spectrum of 1525 to 1559 MHz was never envisioned as being used for high-speed wireless broadband based on the 2003 and 2004 FCC ATC rulings making clear that the Ancillary Tower Component (ATC) would be, in fact, ancillary to the primary satellite component. To build public support of efforts to continue the 2004 FCC authorization of LightSquared's ancillary terrestrial component vs. a simple ground-based LTE service in the Mobile Satellite Service band, GPS receiver manufacturer Trimble Navigation Ltd. formed the "Coalition To Save Our GPS."

The FCC and LightSquared have each made public commitments to solve the GPS interference issue before the network is allowed to operate. However, according to Chris Dancy of the Aircraft Owners and Pilots Association, airline pilots with the type of systems that would be affected "may go off course and not even realize it." The problems could also affect the Federal Aviation Administration upgrade to the air traffic control system, United States Defense Department guidance, and local emergency services including 911.

On February 14, 2012, the U.S. Federal Communications Commission (FCC) moved to bar LightSquared's planned national broadband network after being informed by the National Telecommunications and Information Administration (NTIA), the federal agency that coordinates spectrum uses for the military and other federal government entities, that "there is no practical way to mitigate potential interference at this time".LightSquared is challenging the FCC's action.

SATELLITE NAVIGATION

A satellite navigation or satnav system is a system that uses satellites to provide autonomous geo-spatial positioning. It allows small electronic receivers to determine their location (longitude, latitude, and altitude/elevation) to high precision (within a few meters) using time signals transmitted along a line of sight by radio from satellites. The system can be used for navigation or for tracking the position of something fitted with a receiver (satellite tracking). The signals also allow the electronic receiver to calculate the current local time to high precision, which allows time synchronization. Satnav systems operate independently of any telephonic or internet reception, though these technologies can enhance the usefulness of the positioning information generated.

A satellite navigation system with global coverage may be termed a global navigation satellite system (GNSS). As of December 2016 only the United States NAVSTAR Global Positioning System (GPS), the Russian GLONASS and the EUs Galileo are global operational GNSSs. China is in the process of expanding its regional BeiDou Navigation Satellite System into the global Compass navigation system by 2020. The European Union's Galileo is a global GNSS scheduled to be fully operational by 2020. India currently has satellite-based augmentation system, GPS Aided GEO Augmented Navigation (GAGAN), which enhances the accuracy of NAVSTAR GPS and GLONASS positions. India has already launched the IRNSS, with an operational name NAVIC (Navigation with Indian Constellation), a constellation of satellites for navigation in and around the Indian Subcontinent. It is expected to be fully operational by June 2016. France and Japan are in the process of developing regional navigation systems as well.

Global coverage for each system is generally achieved by a satellite constellation of 20–30 medium Earth orbit (MEO) satellites spread between several orbital planes. The actual systems vary, but use orbital inclinations of >50° and orbital periods of roughly twelve hours (at an altitude of about 20,000 kilometers or 12,000 miles).

Classification

Satellite navigation systems that provide enhanced accuracy and integrity monitoring usable for civil navigation are classified as follows:

- **GNSS-1** is the first generation system and is the combination of existing satellite navigation systems (GPS and GLONASS), with Satellite Based Augmentation Systems (SBAS) or Ground Based Augmentation Systems (GBAS). In the United States, the satellite based component is the Wide Area Augmentation System (WAAS), in Europe it is the European Geostationary Navigation Overlay Service (EGNOS), and in Japan it is the Multi-Functional Satellite Augmentation System (MSAS). Ground based augmentation is provided by systems like the Local Area Augmentation System (LAAS).
- **GNSS-2** is the second generation of systems that independently provides a full civilian satellite navigation system, exemplified by the European Galileo positioning system. These systems will provide the accuracy and integrity monitoring necessary for civil navigation; including aircraft. This system consists of L1 and L2 frequencies (in the L band of the radio spectrum) for civil use and L5 for system integrity. Development is also in progress to provide GPS with civil use L2 and L5 frequencies, making it a GNSS-2 system.[1]
- Core Satellite navigation systems, currently GPS (United States), GLONASS

(Russian Federation), Galileo (European Union) and Compass (China).

- Global Satellite Based Augmentation Systems (SBAS) such as Omnistar and StarFire.
- Regional SBAS including WAAS (US), EGNOS (EU), MSAS (Japan) and GAGAN (India).
- Regional Satellite Navigation Systems such as China›s Beidou, India›s NAVIC, and Japan›s proposed QZSS.
- Continental scale Ground Based Augmentation Systems (GBAS) for example the Australian GRAS and the US Department of Transportation National Differential GPS (DGPS) service.
- Regional scale GBAS such as CORS networks.
- Local GBAS typified by a single GPS reference station operating Real Time Kinematic (RTK) corrections.

History and Theory

Early predecessors were the ground based DECCA, LORAN, GEE and Omega radio navigation systems, which used terrestrial longwave radio transmitters instead of satellites. These positioning systems broadcast a radio pulse from a known "master" location, followed by a pulse repeated from a number of "slave" stations. The delay between the reception of the master signal and the slave signals allowed the receiver to deduce the distance to each of the slaves, providing a fix.

ACCURACY OF NAVIGATION SYSTEMS
(2-dimensional)

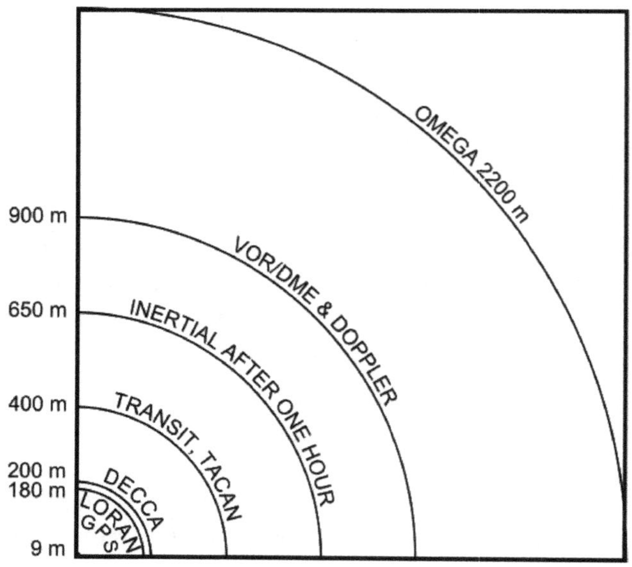

The first satellite navigation system was Transit, a system deployed by the US military in the 1960s. Transit's operation was based on the Doppler effect: the satellites travelled on well-known paths and broadcast their signals on a well-known frequency. The received frequency will differ slightly from the broadcast frequency because of the movement of the satellite with respect to the receiver. By monitoring this frequency shift over a short time interval, the receiver can determine its location to one side or the other of the satellite, and several such measurements combined with a precise knowledge of the satellite's orbit can fix a particular position.

Part of an orbiting satellite's broadcast included its precise orbital data. In order to ensure accuracy, the US Naval Observatory (USNO) continuously observed the precise orbits of these satellites. As a satellite's orbit deviated, the USNO would send the updated information to the satellite. Subsequent broadcasts from an updated satellite would contain the most recent accurate information about its orbit.

Modern systems are more direct. The satellite broadcasts a signal that contains orbital data (from which the position of the satellite can be calculated) and the precise time the signal was transmitted. The orbital data is transmitted in a data message that is superimposed on a code that serves as a timing reference. The satellite uses an atomic clock to maintain synchronization of all the satellites in the constellation. The receiver compares the time of broadcast encoded in the transmission of three (at sea level) or four different satellites, thereby measuring the time-of-flight to each satellite. Several such measurements can be made at the same time to different satellites, allowing a continual fix to be generated in real time using an adapted version of trilateration.

Each distance measurement, regardless of the system being used, places the receiver on a spherical shell at the measured distance from the broadcaster. By taking several such measurements and then looking for a point where they meet, a fix is generated. However, in the case of fast-moving receivers, the position of the signal moves as signals are received from several satellites. In addition, the radio signals slow slightly as they pass through the ionosphere, and this slowing varies with the receiver's angle to the satellite, because that changes the distance through the ionosphere. The basic computation thus attempts to find the shortest directed line tangent to four oblate spherical shells centred on four satellites. Satellite navigation receivers reduce errors by using combinations of signals from multiple satellites and multiple correlators, and then using techniques such as Kalman filtering to combine the noisy, partial, and constantly changing data into a single estimate for position, time, and velocity.

Civil and Military Uses

The original motivation for satellite navigation was for military applications. Satellite navigation allows the precision in the delivery of weapons to targets, greatly increasing their lethality whilst reducing inadvertent casualties from mis-directed weapons. Satellite navigation also allows forces to be directed and to locate themselves more easily, reducing the fog of war.

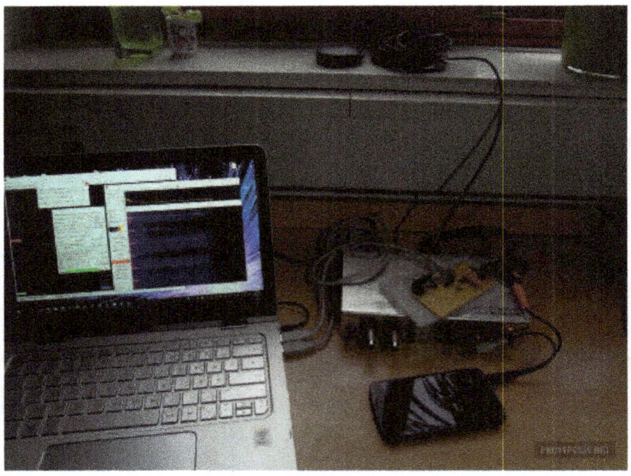

Satellite navigation using a laptop and a GPS receiver

The ability to supply satellite navigation signals is also the ability to deny their availability. The operator of a satellite navigation system potentially has the ability to degrade or eliminate satellite navigation services over any territory it desires.

Global Satellite Navigation Systems

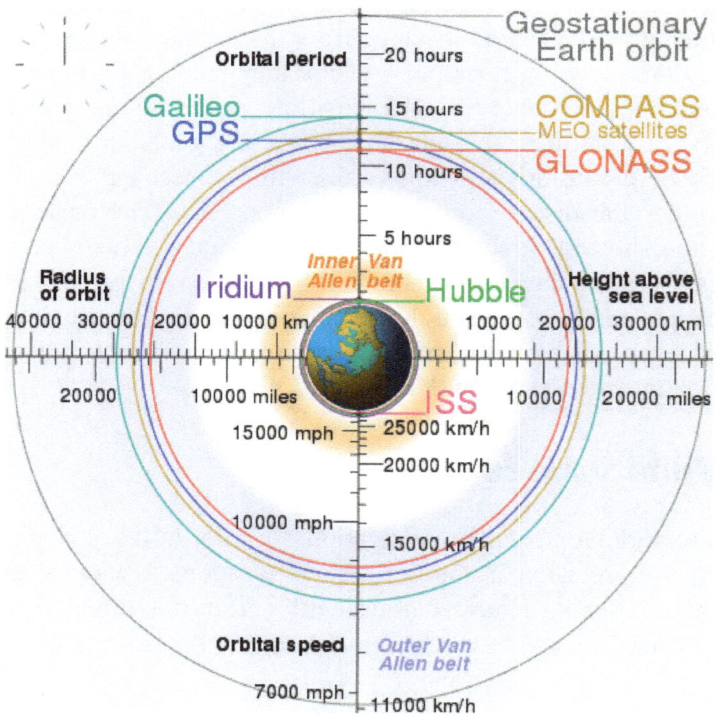

Comparison of geostationary, **GPS**, **GLONASS**, **Galileo**, **Compass (MEO)**, International Space Station, Hubble Space Telescope and Iridium constellation orbits, with the Van Allen radiation belts and the Earth to scale. The Moon's orbit is around 9 times larger than geostationary orbit.

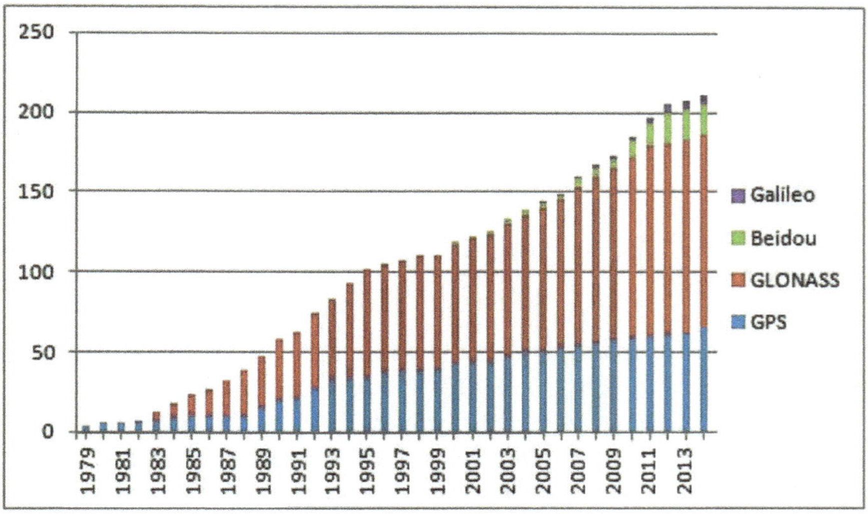

launched GNSS satellites 1978 to 2014

Operational

GPS

The United States' Global Positioning System (GPS) consists of up to 32 medium Earth orbit satellites in six different orbital planes, with the exact number of satellites varying as older satellites are retired and replaced. Operational since 1978 and globally available since 1994, GPS is currently the world's most utilized satellite navigation system.

GLONASS

The formerly Soviet, and now Russian, ***Glo****bal'naya* ***Na****vigatsionnaya* ***S****putnikovaya* ***S****istema* (Russian: ГЛОбальная НАвигационная Спутниковая Система, GLObal NAvigation Satellite System), or GLONASS, is a space-based satellite navigation system that provides a civilian radionavigation-satellite service and is also used by the Russian Aerospace Defence Forces. The full orbital constellation of 24 GLONASS satellites enables full global coverage.

Galileo

The European Union and European Space Agency agreed in March 2002 to introduce their own alternative to GPS, called the Galileo positioning system. Galileo became operational on 15 December 2016. At an estimated cost of EUR 3.0 billion, the system of 30 MEO satellites was originally scheduled to be operational in 2010. The original year to become operational was 2014. The first experimental satellite was launched on 28 December 2005. Galileo is expected to be compatible with the modernized GPS system.

The receivers will be able to combine the signals from both Galileo and GPS satellites to greatly increase the accuracy. Galileo is expected to be in full service in 2020 and at a substantially higher cost. The main modulation used in Galileo Open Service signal is the Composite Binary Offset Carrier (CBOC) modulation.

In Development

BeiDou

China has indicated they plan to complete the entire second generation Beidou Navigation Satellite System (BDS or BeiDou-2, formerly known as COMPASS), by expanding current regional (Asia-Pacific) service into global coverage by 2020. The BeiDou-2 system is proposed to consist of 30 MEO satellites and five geostationary satellites. A 16-satellite regional version (covering Asia and Pacific area) was completed by December 2012.

Comparison of Systems

System	**GPS**	**GLONASS**	**BeiDou**	**Galileo**	**NAVIC**
Owner	United States	Russian Federation	China	European Union	India
Coding	CDMA	FDMA	CDMA	CDMA	CDMA
Orbital altitude	20,180 km (12,540 mi)	19,130 km (11,890 mi)	21,150 km (13,140 mi)	23,222 km (14,429 mi)	36,000 km (22,000 mi)
Period	11.97 h (11 h 58 min)	11.26 h (11 h 16 min)	12.63 h (12 h 38 min)	14.08 h (14 h 5 min)	
Revolutions per sidereal day	2	17/8	17/9	17/10	
Number of satellites	**32** (at least 24 by design)	**28** (at least 24 by design) including: 24 operational 2 under check by the satellite prime contractor 2 in flight tests phase	5 geostationary orbit (GEO) satellites, 30 medium Earth orbit (MEO) satellites	18 full operation capable satellites in orbit 30 operational satellites budgeted	Total : 7 In Orbit : 7
Frequency	1.57542 GHz (L1 signal) 1.2276 GHz (L2 signal)	Around 1.602 GHz (SP) Around 1.246 GHz (SP)	1.561098 GHz (B1) 1.589742 GHz (B1-2) 1.20714 GHz (B2) 1.26852 GHz (B3)	1.164–1.215 GHz (E5a and E5b) 1.260–1.300 GHz (E6) 1.559–1.592 GHz (E2-L1-E11)	L5-band 1164.45–1188.45 MHz S-band 2483.5-2500 MHz)
Status	Operational	Operational	22 satellites operational, 40 additional satellites 2016-2020	18 satellites operational, 12 additional satellites 2016-2020	Operational

Regional Satellite Navigation Systems

BeiDou-1

Chinese regional (Asia-Pacific, 16 satellites) network to be expanded into the whole global system which consists of all 35 satellites by 2020.

NAVIC

The NAVIC or NAVigation with Indian Constellation is an autonomous regional satellite navigation system developed by Indian Space Research Organisation (ISRO) which would be under the total control of Indian government. The government approved the project in May 2006, with the intention of the system completed and implemented on 28 April 2016. It will consist of a constellation of 7 navigational satellites. 3 of the satellites will be placed in the Geostationary orbit (GEO) and the remaining 4 in the Geosynchronous orbit(GSO) to have a larger signal footprint and lower number of satellites to map the region. It is intended to provide an all-weather absolute position accuracy of better than 7.6 meters throughout India and within a region extending approximately 1,500 km around it. A goal of complete Indian control has been stated, with the space segment, ground segment and user receivers all being built in India. All seven satellites, IRNSS-1A, IRNSS-1B, IRNSS-1C, IRNSS-1D, IRNSS-1E, IRNSS-1F, and IRNSS-1G, of the proposed constellation were precisely launched on 1 July 2013, 4 April 2014, 16 October 2014, 28 March 2015, 20 January 2016, 10 March 2016 and 28 April 2016 respectively from Satish Dhawan Space Centre. The system is expected to be fully operational by August 2016.

QZSS

The Quasi-Zenith Satellite System (QZSS), is a proposed three-satellite regional time transfer system and enhancement for GPS covering Japan. The first demonstration satellite was launched in September 2010.

Augmentation

GNSS augmentation is a method of improving a navigation system's attributes, such as accuracy, reliability, and availability, through the integration of external information into the calculation process, for example, the Wide Area Augmentation System, the European Geostationary Navigation Overlay Service, the Multi-functional Satellite Augmentation System, Differential GPS, and inertial navigation systems.

DORIS

Doppler Orbitography and Radio-positioning Integrated by Satellite (DORIS) is a French precision navigation system. Unlike other GNSS systems, it is based on static emitting stations around the world, the receivers being on satellites, in order to precisely determine their orbital position. The system may be used also for mobile receivers on land with more limited usage and coverage. Used with traditional GNSS systems, it pushes the accuracy of positions to centimetric precision (and to millimetric precision for altimetric application and also allows monitoring very tiny seasonal changes of Earth rotation and deformations), in order to build a much more precise geodesic reference system.

Low Earth Orbit Satellite Phone Networks

The two current operational low Earth orbit satellite phone networks are able to track transceiver units with accuracy of a few kilometers using doppler shift calculations from the satellite. The coordinates are sent back to the transceiver unit where they can be read using AT commands or a graphical user interface. This can also be used by the gateway to enforce restrictions on geographically bound calling plans.

REFERENCES

1. "Navstar GPS Space Segment/Navigation User Interfaces"
2. "The Navstar Global Positioning System, hereafter referred to as GPS, is a space-based radionavigation system owned by the United States Government (USG) and operated by the United States Air Force (USAF)."
3. "Wide Area Augmentation System (WAAS) Performance Standard", Section B.3, Abbreviations and Acronyms: "GPS: Global Positioning System (or Navstar Global Positioning System)"
4. http://www.space.com/19794-navstar.html
5. "What is a GPS?".
6. Srivastava, Ishan (5 April 2014). "How Kargil spurred India to design own GPS". The Times of India. Retrieved 9 December 2014.
7. National Research Council (U.S.). Committee on the Future of the Global Positioning System; National Academy of Public Administration (1995). The global positioning system: a shared national asset: recommendations for technical improvements and enhancements. National Academies Press. p. 16. ISBN 0-309-05283-1. Retrieved August 16, 2013., https://books.google.com/books?id=FAHk65slfY4C&pg=PA16
8. O'Leary, Beth Laura; Darrin, Ann Garrison (2009). Handbook of Space Engineering, Archaeology, and Heritage. Hoboken: CRC Press. pp. 239–240. ISBN 9781420084320.
9. "Factsheets : GPS Advanced Control Segment (OCX)". Losangeles.af.mil. October 25, 2011. Retrieved November 6, 2011.
10. "Russia Launches Three More GLONASS-M Space Vehicles". Inside GNSS. Retrieved December 26, 2008.
11. GLONASS the future for all smartphones?

12. Winterberg, Friedwardt (1956). "Relativistische Zeitdiiatation eines künstlichen Satelliten (Relativistic time dilation of an artificial satellite)". Astronautica Acta II (in German) (25). Retrieved 19 October 2014.
13. "GPS and Relativity". Astronomy.ohio-state.edu. Retrieved November 6, 2011.
14. Guier, William H.; Weiffenbach, George C. (1997). "Genesis of Satellite Navigation" (PDF). Johns Hopkins APL Technical Digest. 19 (1): 178–181.
15. Steven Johnson (2010), Where good ideas come from, the natural history of innovation, New York: Riverhead Books
16. Helen E. Worth; Mame Warren (2009). Transit to Tomorrow. Fifty Years of Space Research at The Johns Hopkins University Applied Physics Laboratory (PDF).
17. Catherine Alexandrow (April 2008). "The Story of GPS".
18. DARPA: 50 Years of Bridging the Gap. April 2008.
19. Howell, Elizabeth. "Navstar: GPS Satellite Network". SPACE.com. Retrieved February 14, 2013.
20. Jerry Proc. "Omega". Jproc.ca. Retrieved December 8, 2009.
21. "Why Did the Department of Defense Develop GPS?". Trimble Navigation Ltd. Archived from the original on October 18, 2007. Retrieved January 13, 2010.
22. "Charting a Course Toward Global Navigation". The Aerospace Corporation. Archived from the original on November 1, 2002. Retrieved October 14, 2013.
23. "A Guide to the Global Positioning System (GPS) — GPS Timeline". Radio Shack. Retrieved January 14, 2010.
24. "GEODETIC EXPLORER-A Press Kit" (PDF). NASA. October 29, 1965. Retrieved 20 October 2015.
25. "SECOR Chronology". Mark Wade's Encyclopedia Astronautica. Retrieved January 19, 2010.
26. "MX Deployment Reconsidered." Retrieved: 7 June 2013.
27. Michael Russell Rip; James M. Hasik (2002). The Precision Revolution: GPS and the Future of Aerial Warfare. Naval Institute Press. p. 65. ISBN 1-55750-973-5. Retrieved January 14, 2010.
28. Hegarty, Christopher J.; Chatre, Eric (December 2008). "Evolution of the Global Navigation SatelliteSystem (GNSS)". Proceedings of the IEEE: 1902–1917. doi:10.1109/JPROC.2008.2006090.
29. "ICAO Completes Fact-Finding Investigation". International Civil Aviation Organization. Archived from the original on May 17, 2008. Retrieved September 15, 2008.
30. "United States Updates Global Positioning System Technology". America.gov. February 3, 2006.
31. Rumerman, Judy A. (2009). NASA Historical Data Book, Volume VII (PDF). NASA. p. 136.

32. The Global Positioning System Assessing National Policies, by Scott Pace, Gerald P. Frost, Irving Lachow, David R. Frelinger, Donna Fossum, Don Wassem, Monica M. Pinto, Rand Corporation, 1995,Appendix B, GPS History, Chronology, and Budgets
33. "GPS & Selective Availability Q&A" (PDF). NOAA]. Archived from the original (PDF) on September 21, 2005. Retrieved May 28, 2010.
34. "GPS Accuracy". GPS.gov. GPS.gov. Retrieved 4 May 2015.
35. E. Steitz, David. "NATIONAL POSITIONING, NAVIGATION AND TIMING ADVISORY BOARD NAMED". Retrieved March 22, 2007.
36. GPS Wing Reaches GPS III IBR Milestone in Inside GNSS November 10, 2008
37. GPS CONSTELLATION STATUS FOR 08/26/2015
38. "Recap story: Three Atlas 5 launch successes in one month".
39. "GPS almanacs". Navcen.uscg.gov. Retrieved October 15, 2010.
40. Dietrich Schroeer; Mirco Elena (2000). Technology Transfer. Ashgate. p. 80. ISBN 0-7546-2045-X. Retrieved May 25, 2008.
41. Michael Russell Rip; James M. Hasik (2002). The Precision Revolution: GPS and the Future of Aerial Warfare. Naval Institute Press. ISBN 1-55750-973-5. Retrieved May 25, 2008.
42. "AF Space Command Chronology". USAF Space Command. Retrieved June 20, 2011.
43. "FactSheet: 2nd Space Operations Squadron". USAF Space Command. Retrieved June 20, 2011.
44. The Global Positioning System: Assessing National Policies, p.245. RAND corporation
45. "USNO NAVSTAR Global Positioning System". U.S. Naval Observatory. Retrieved January 7, 2011.
46. National Archives and Records Administration. U.S. Global Positioning System Policy. March 29, 1996.
47. "National Executive Committee for Space-Based Positioning, Navigation, and Timing". Pnt.gov. Archived from the original on May 28, 2010. Retrieved October 15, 2010.
48. "Assisted-GPS Test Calls for 3G WCDMA Networks". 3g.co.uk. November 10, 2004. Retrieved November 24, 2010.
49. "First Modernized GPS Satellite Built By Lockheed Martin Launched". Phys. org. Retrieved September 26, 2005.
50. 010907 (September 17, 2007). "losangeles.af.mil". losangeles.af.mil. Retrieved October 15, 2010.
51. Johnson, Bobbie (May 19, 2009). "GPS system 'close to breakdown'". The Guardian. London. Retrieved December 8, 2009.

52. Coursey, David (May 21, 2009). "Air Force Responds to GPS Outage Concerns". ABC News. Retrieved May 22, 2009.

53. "Air Force GPS Problem: Glitch Shows How Much U.S. Military Relies On GPS". Huffingtonpost.comm. June 1, 2010. Retrieved October 15, 2010.

54. "Contract Award for Next Generation GPS Control Segment Announced". Retrieved December 14, 2012.

55. United States Naval Research Laboratory. National Medal of Technology for GPS. November 21, 2005

56. "Space Technology Hall of Fame, Inducted Technology: Global Positioning System (GPS)". Archived from the original on June 12, 2012.

57. Abel, J.S. and Chaffee, J.W., "Existence and uniqueness of GPS solutions", IEEE Transactions on Aerospace and Electronic Systems, vol:26, no:6, p:748-53, Sept. 1991.

58. Fang, B.T., "Comments on "Existence and uniqueness of GPS solutions" by J.S. Abel and J.W. Chaffee", IEEE Transactions on Aerospace and Electronic Systems, vol:28 , no:4, Oct. 1992.

59. Grewal, Mohinder S.; Weill, Lawrence R.; Andrews, Angus P. (2007). Global Positioning Systems, Inertial Navigation, and Integration (2nd ed.). John Wiley & Sons. pp. 92–93. ISBN 0-470-09971-2., https://books.google.com/books?id=6P7UNphJ1z8C&pg=PA92

60. Georg zur Bonsen; Daniel Ammann; Michael Ammann; Etienne Favey; Pascal Flammant (April 1, 2005). "Continuous Navigation Combining GPS with Sensor-Based Dead Reckoning". GPS World. Archived from the original on November 11, 2006.

61. "NAVSTAR GPS User Equipment Introduction" (PDF). United States Government. Chapter 7

62. "GPS Support Notes" (PDF). January 19, 2007. Archived from the original (PDF) on March 27, 2009. Retrieved November 10, 2008.

63. John Pike. "GPS III Operational Control Segment (OCX)". Globalsecurity.org. Retrieved December 8, 2009.

64. "Global Positioning System". Gps.gov. Archived from the original on July 30, 2010. Retrieved June 26, 2010.

65. Daly, P. "Navstar GPS and GLONASS: global satellite navigation systems" (PDF). IEEE.

66. Dana, Peter H. (August 8, 1996). "GPS Orbital Planes" (GIF).

67. GPS Overview from the NAVSTAR Joint Program Office. Retrieved December 15, 2006.

68. What the Global Positioning System Tells Us about Relativity Archived January 4, 2007, at the Wayback Machine.. Retrieved January 2, 2007.

69. "Archived copy". Archived from the original on October 22, 2011. Retrieved 2011-10-27.. Retrieved October 27, 2011

70. "USCG Navcen: GPS Frequently Asked Questions". Retrieved January 31, 2007.
71. Thomassen, Keith. "How GPS Works". avionicswest.com. Archived from the original on March 30, 2016. Retrieved April 22, 2014.
72. Samama, Nel (2008). Global Positioning: Technologies and Performance. John Wiley & Sons. p. 65. ISBN 0-470-24190-X., https://books.google.com/books?id=EyFrcnSRFFgC&pg=PA65
73. Agnew, D.C.; Larson, K.M. (2007). "Finding the repeat times of the GPS constellation". GPS Solutions. Springer. 11 (1): 71–76. doi:10.1007/s10291-006-0038-4. This article from author's web site Archived February 16, 2008, at the Wayback Machine., with minor correction.
74. "CURRENT GPS CONSTELLATION". U.S. Naval Observatory.
75. Massatt, Paul; Wayne Brady (Summer 2002). "Optimizing performance through constellation management" (PDF). Crosslink: 17–21. Archived from the original on January 25, 2012.
76. United States Coast Guard General GPS News 9–9–05
77. USNO NAVSTAR Global Positioning System. Retrieved May 14, 2006.
78. "GPS III Operational Control Segment (OCX)". GlobalSecurity.org.
79. "The USA's GPS-III Satellites". Defense Industry Daily. October 13, 2011.
80. "GPS Completes Next Generation Operational Control System PDR". Air Force Space Command News Service. September 14, 2011.
81. "'Embarrassing to defend': US general blasts Raytheon's GPS control system a 'disaster'". RT. December 9, 2015.
82. "Publications and Standards from the National Marine Electronics Association (NMEA)". National Marine Electronics Association. Retrieved June 27, 2008.
83. Exploring GPS Data for Operational Analysis of Farm Machinery
84. A lecture note on Global Positioning System in Precision Agriculture
85. "Common View GPS Time Transfer". nist.gov. Archived from the original on October 28, 2012. Retrieved July 23, 2011.
86. "Using GPS to improve tropical cyclone forecasts". ucar.edu.
87. "Spotlight GPS pet locator". Spotlightgps.com. Retrieved October 15, 2010.
88. Khetarpaul, S., Chauhan, R., Gupta, S. K., Subramaniam, L. V., Nambiar, U. (2011). Mining GPS data to determine interesting locations. Proceedings of the 8th International Workshop on Information Integration on the Web.
89. "The Use of GPS Tracking Technology in Australian Football". Retrieved 2016-09-25.
90. "The Pacific Northwest Geodetic Array". cwu.edu.
91. Arms Control Association.Missile Technology Control Regime. Retrieved May 17, 2006.
92. Sinha, Vandana (July 24, 2003). "Commanders and Soldiers' GPS-receivers". Gcn.com. Retrieved October 13, 2009.

93. "XM982 Excalibur Precision Guided Extended Range Artillery Projectile". GlobalSecurity.org. May 29, 2007. Retrieved September 26, 2007. (registration required (help)).

94. Sandia National Laboratory's Nonproliferation programs and arms control technology.

95. Dennis D. McCrady. "The GPS Burst Detector W-Sensor" (PDF). Sandia National Laboratories.

96. "US Air Force Eyes Changes To National Security Satellite Programs.". Aviationweek.com. January 18, 2013. Retrieved September 28, 2013.

97. Greenemeier, Larry. "GPS and the World's First "Space War"". Scientific American. Retrieved 2016-02-08.

98. "Satellite message format". Gpsinformation.net. Retrieved October 15, 2010.

99. Peter H. Dana. "GPS Week Number Rollover Issues". Retrieved August 12, 2013.

100. "Interface Specification IS-GPS-200, Revision D: Navstar GPS Space Segment/ Navigation User Interfaces" (PDF). Navstar GPS Joint Program Office. p. 103.

101. Richharia, Madhavendra; Westbrook, Leslie David (2011). Satellite Systems for Personal Applications: Concepts and Technology. John Wiley & Sons. p. 443. ISBN 1-119-95610-2., https://books.google.com/books?id=MqPQ5CbgQ48C443

102. Misra, Pratap; Enge, Per (2006). Global Positioning System. Signals, Measurements and Performance (2nd ed.). Ganga-Jamuna Press. p. 115. ISBN 0-9709544-1-7. Retrieved August 16, 2013.

103. Borre, Kai; M. Akos, Dennis; Bertelsen, Nicolaj; Rinder, Peter; Jensen, Søren Holdt (2007). A Software-Defined GPS and Galileo Receiver. A single-Frequency Approach. Springer. p. 18. ISBN 0-8176-4390-7.

104. TextGenerator Version 2.0. "United States Nuclear Detonation Detection System (USNDS)". Fas.org. Retrieved November 6, 2011.

105. "First Block 2F GPS Satellite Launched, Needed to Prevent System Failure". DailyTech. Retrieved May 30, 2010.

106. "Air Force Successfully Transmits an L5 Signal From GPS IIR-20(M) Satellite". LA AFB News Release. Retrieved June 20, 2011.

107. "Federal Communications Commission Presented Evidence of GPS Signal Interference". GPS World. Archived from the original on October 11, 2011. Retrieved November 6, 2011.

108. "Coalition to Save Our GPS". Saveourgps.org. Retrieved November 6, 2011.

109. "LightSquared Tests Confirm GPS Jamming". Aviation Week. Archived from the original on August 12, 2011. Retrieved June 20, 2011.

110. "GPS Almanacs, NANUS, and Ops Advisories (including archives)". GPS Almanac Information. United States Coast Guard. Retrieved September 9, 2009.

111. "George, M., Hamid, M., and Miller A. Gold Code Generators in Virtex Devices at the Internet ArchivePDF

112. section 4 beginning on page 15 GEOFFREY BLEWITT: BASICS OF THE GPS TECHNIQUE
113. "Global Positioning Systems" (PDF). Archived from the original (PDF) on July 19, 2011. Retrieved October 15, 2010.
114. Dana, Peter H. "Geometric Dilution of Precision (GDOP) and Visibility". University of Colorado at Boulder. Retrieved July 7, 2008.
115. Peter H. Dana. "Receiver Position, Velocity, and Time". University of Colorado at Boulder. Retrieved July 7, 2008.
116. "The Mathematics of GPS". siam.org. Archived from the original on March 16, 2005.
117. "GPS Solutions: Closed Forms, Critical and Special Configurations of P4P".
118. Bancroft, S. (January 1985). "An Algebraic Solution of the GPS Equations". IEEE Transactions on Aerospace and Electronic Systems. AES-21: 56–59. Bibcode:1985ITAES..21...56B. doi:10.1109/TAES.1985.310538.[dead link]
119. Chaffee, J. and Abel, J., "On the Exact Solutions of Pseudorange Equations", IEEE Transactions on Aerospace and Electronic Systems, vol:30, no:4, pp: 1021–1030, 1994
120. Sirola, Niilo, "Closed-form Algorithms in Mobile Positioning: Myths and Misconceptions", Proceedings of the 7th Workshop on Positioning, Navigation and Communication 2010 (WPNC'10), Dresden, March 2010. Retrieved November 10, 2014
121. "GNSS Positioning Approaches – GPS Satellite Surveying, Fourth Edition – Leick – Wiley Online Library". wiley.com.
122. Alfred Kleusberg, "Analytical GPS Navigation Solution", University of Stuttgart Research Compendium,1994
123. Oszczak, B., "New Algorithm for GNSS Positioning Using System of Linear Equations," Proceedings of the 26th International Technical Meeting of The Satellite Division of the Institute of Navigation (ION GNSS+ 2013), Nashville, TN, September 2013, pp. 3560–3563.
124. Attewill, Fred. (2013-02-13) Vehicles that use GPS jammers are big threat to aircraft. Metro.co.uk. Retrieved on 2013-08-02.
125. "Frequently Asked Questions About Selective Availability". National Coordination Office for Space-Based Positioning, Navigation, and Timing (PNT). October 2001. Retrieved 2015-06-13. Selective Availability ended a few minutes past midnight EDT after the end of May 1, 2000. The change occurred simultaneously across the entire satellite constellation.
126. https://blackboard.vuw.ac.nz/bbcswebdav/pid-1444805-dt-content-rid-2193398_1/courses/2014.1.ESCI203/Esci203_2014_GPS_1.pdf (subscription required)
127. McNamara, Joel (2008). GPS For Dummies. John Wiley & Sons. p. 59. ISBN 0-470-45785-6., https://books.google.com/books?id=Hbz4LYIrvuMC&pg=PA59
128. "NAVSTAR GPS User Equipment Introduction" (PDF). Section 1.2.2

129. "NOTICE ADVISORY TO NAVSTAR USERS (NANU) 2015086". Retrieved October 2015. Check date values in: |access-date= (help)
130. "Notice Advisory to Navstar Users (NANU) 2012034". GPS Operations Center. May 30, 2012. Retrieved July 2, 2012.
131. David W. Allan (1997). "The Science of Timekeeping" (PDF). Hewlett Packard. Archived (PDF) from the original on October 12, 2012.
132. "The Role of GPS in Precise Time and Frequency Dissemination" (PDF). GPSworld. July–August 1990. Retrieved April 27, 2014.
133. "GPS time accurate to 100 nanoseconds". Galleon. Retrieved October 12, 2012.
134. "Between-Satellite Differencing". Gmat.unsw.edu.au. Retrieved October 15, 2010.
135. "Double differencing". Gmat.unsw.edu.au. Retrieved October 15, 2010.
136. "Triple differencing". Gmat.unsw.edu.au. Retrieved October 15, 2010.
137. chapter on root finding and nonlinear sets of equations
138. Preview of Root Finding. Books.google.com. 2007. ISBN 978-0-521-88068-8. Retrieved October 15, 2010.
139. "2011 John Deere StarFire 3000 Operator Manual" (PDF). John Deere. Archived from the original (PDF) on January 5, 2012. Retrieved November 13, 2011.
140. "Federal Communications Commission Report and Order In the Matter of Fixed and Mobile Services in the Mobile Satellite Service Bands at 1525–1559 MHz and 1626.5–1660.5 MHz" (PDF). FCC.gov. April 6, 2011. Retrieved December 13, 2011.
141. "Federal Communications Commission Table of Frequency Allocations" (PDF). FCC.gov. November 18, 2011. Retrieved December 13, 2011.
142. "FCC Docket File Number: SATASG20010302000017, "Mobile Satellite Ventures LLC Application for Assignment and Modification of Licenses and for Authority to Launch and Operate a Next-Generation Mobile Satellite System"". FCC.gov. March 1, 2001. p. 9.
143. "U.S. GPS Industry Council Petition to the FCC to adopt OOBE limits jointly proposed by MSV and the Industry Council". FCC.gov. September 4, 2003. Retrieved December 13, 2011.
144. "ORDER ON RECONSIDERATION" (PDF). Jul 3, 2003. Retrieved October 2015. Check date values in: |access-date= (help)
145. "Statement of Julius P. Knapp, Chief, Office of Engineering and Technology, Federal Communications Commission" (PDF). gps.gov. September 15, 2011. p. 3. Retrieved December 13, 2011.
146. "FCC Order, Granted LightSquared Subsidiary LLC, a Mobile Satellite Service licensee in the L-Band, a conditional waiver of the Ancillary Terrestrial Component "integrated service" rule" (PDF). Federal Communications Commission. FCC. Gov. January 26, 2011. Retrieved December 13, 2011.

147. "Data Shows Disastrous GPS Jamming from FCC-Approved Broadcaster". gpsworld.com. February 1, 2011. Archived from the original on February 6, 2011. Retrieved February 10, 2011.

148. "Javad Ashjaee GPS World webinar". gpsworld.com. December 8, 2011. Archived from the original on November 26, 2011. Retrieved December 13, 2011.

149. "FCC Order permitting mobile satellite services providers to provide an ancillary terrestrial component (ATC) to their satellite systems" (PDF). Federal Communications Commission. FCC.gov. February 10, 2003. Retrieved December 13, 2011.

150. "Federal Communications Commission Fixed and Mobile Services in the Mobile Satellite Service". Federal Communications Commission. FCC.gov. July 15, 2010. Retrieved December 13, 2011.

151. Archived December 13, 2012, at the Wayback Machine.

152. "Coalition to Save Our GPS". Saveourgps.org. Retrieved November 6, 2011.

153. Jeff Carlisle (June 23, 2011). "Testimony of Jeff Carlisle, LightSquared Executive Vice President of Regulatory Affairs and Public Policy to U.S. House Subcommittee on Aviation and Subcommittee on Coast Guard and Maritime Transportation" (PDF). Archived from the original (PDF) on September 29, 2011. Retrieved December 13, 2011.

154. Julius Genachowski (May 31, 2011). "FCC Chairman Genachowski Letter to Senator Charles Grassley" (PDF). Retrieved December 13, 2011.

155. Tessler, Joelle (April 7, 2011). "Internet network may jam GPS in cars, jets". The Sun News. Archived from the original on May 1, 2011. Retrieved April 7, 2011.

156. FCC press release "Spokesperson Statement on NTIA Letter – LightSquared and GPS". February 14, 2012. Accessed 2013-03-03.

157. Paul Riegler, FBT. "FCC Bars LightSquared Broadband Network Plan". February 14, 2012. Retrieved February 14, 2012.

158. "Galileo navigation satellite system goes live". dw.com. Retrieved December 17, 2016.

159. Beidou coverage

160. "Beidou satellite navigation system to cover whole world in 2020". Eng.chinamil.com.cn. Retrieved October 15, 2010.

161. Levin, Dan (March 23, 2009). "Chinese Square Off With Europe in Space". The New York Times. China. Retrieved November 6, 2011.

CHAPTER 5

Hyperspectral Imaging

Chapter Contents

- **Introduction**
- **Hyperspectral Image Sensors**
- **Technologies for Hyperspectral Data Acquisition**
- **Distinguishing Hyperspectral from Multispectral Imaging**

Hyperspectral imaging (HSI) combines standard digital imaging with common spectroscopic methods, such as Near-IR, Visible and Fluorescence imaging, to provide increased sensitivity and discrimination capabilities over traditional imaging and detection methods. HSI will not compromise the integrity of your samples, allowing for the possibility of further sample testing, if necessary.

INTRODUCTION

Hyperspectral imaging, like other spectral imaging, collects and processes information from across the electromagnetic spectrum. The goal of hyperspectral imaging is to obtain the spectrum for each pixel in the image of a scene, with the purpose of finding objects, identifying materials, or detecting processes. Nowadays there are two branches of spectral imaging Push broom scanner, which read in an image over time and Snapshot hyperspectral imaging which generates an image in an instance.

Whereas the human eye sees color of visible light in mostly three bands (red, green, and blue), spectral imaging divides the spectrum into many more bands. This technique of dividing images into bands can be extended beyond the visible. In hyperspectral imaging, the recorded spectra have fine wavelength resolution and cover a wide range of wavelengths.

Engineers build hyperspectral sensors and processing systems for applications in astronomy, agriculture, biomedical imaging, geosciences, physics, and surveillance. Hyperspectral sensors look at objects using a vast portion of the electromagnetic spectrum. Certain objects leave unique 'fingerprints' in the electromagnetic spectrum. Known as spectral signatures, these 'fingerprints' enable identification of the materials that make up a scanned object. For example, a spectral signature for oil helps geologists find new oil fields.

Two-dimensional projection of a hyperspectral cube.

HYPERSPECTRAL IMAGE SENSORS

Figuratively speaking, hyperspectral sensors collect information as a set of 'images'. Each image represents a narrow wavelength range of the electromagnetic spectrum, also known as a spectral band. These 'images' are combined to form a three-dimensional (x,y,λ) hyperspectral data cube for processing and analysis, where x and y represent two spatial dimensions of the scene, and λ represents the spectral dimension (comprising a range of wavelengths).

Technically speaking, there are four ways for sensors to sample the hyperspectral cube: Spatial scanning, spectral scanning, snapshot imaging, and spatio-spectral scanning.

Hyperspectral cubes are generated from airborne sensors like the NASA›s Airborne Visible/Infrared Imaging Spectrometer (AVIRIS), or from satellites like NASA's EO-1 with its hyperspectral instrument Hyperion. However, for many development and validation studies, handheld sensors are used.

The precision of these sensors is typically measured in spectral resolution, which is the width of each band of the spectrum that is captured. If the scanner detects a large number of fairly narrow frequency bands, it is possible to identify objects even if they are only captured in a handful of pixels. However, spatial resolution is a factor in addition to spectral resolution. If the pixels are too large, then multiple objects are captured in the same pixel and become difficult to identify. If the pixels are too small, then the energy captured by each sensor cell is low, and the decreased signal-to-noise ratio reduces the reliability of measured features.

The acquisition and processing of hyperspectral images is also referred to as imaging spectroscopy or, with reference to the hyperspectral cube, as 3D spectroscopy.

TECHNOLOGIES FOR HYPERSPECTRAL DATA ACQUISITION

There are four basic techniques for acquiring the three-dimensional (x,y,λ) dataset of a hyperspectral cube. The choice of technique depends on the specific application, seeing that each technique has context-dependent advantages and disadvantages.

Photos illustrating individual sensor outputs for the four hyperspectral imaging techniques. From left to right: Slit spectrum; monochromatic spatial map; 'perspective projection' of hyperspectral cube; wavelength-coded spatial map.

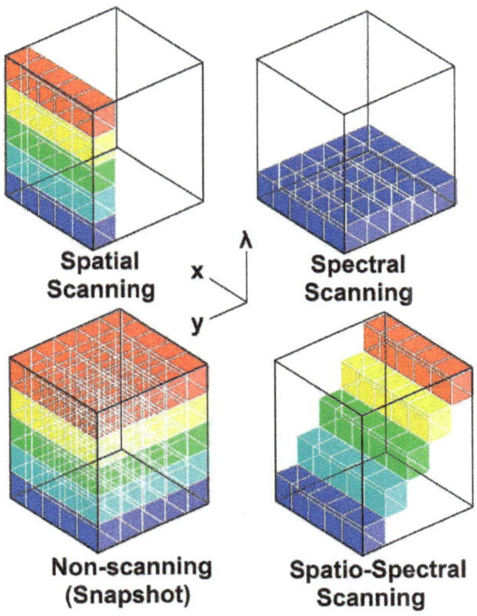

Acquisition techniques for hyperspectral imaging, visualized as sections of the hyperspectral datacube with its two spatial dimensions (x,y) and one spectral dimension (lambda).

Spatial Scanning

In spatial scanning, each two-dimensional (2-D) sensor output represents a full slit spectrum (x,λ). Hyperspectral imaging (HSI) devices for spatial scanning obtain slit spectra by projecting a strip of the scene onto a slit and dispersing the slit image with a prism or a grating. These systems have the drawback of having the image analyzed per lines (with a push broom scanner) and also having some mechanical parts integrated into the optical train. With these line-scan systems, the spatial dimension is collected through platform movement or scanning. This requires stabilized mounts or accurate pointing information to 'reconstruct' the image. Nonetheless, line-scan systems are particularly common in remote sensing, where it is sensible to use mobile platforms. Line-scan systems are also used to scan materials moving by on a conveyor belt. A special case of line scanning is point scanning (with a whisk broom scanner), where a point-like aperture is used instead of a slit, and the sensor is essentially one-dimensional instead of 2-D.

Spectral Scanning

In spectral scanning, each 2-D sensor output represents a monochromatic ('single-colored'), spatial (x,y) map of the scene. HSI devices for spectral scanning are typically based on optical band-pass filters (either tuneable or fixed). The scene is spectrally scanned by exchanging one filter after another while the platform must be stationary. In such 'staring', wavelength scanning systems, spectral smearing can occur if there is movement within the scene, invalidating spectral correlation/detection. Nonetheless, there is the advantage of being able to pick and choose spectral bands, and having a direct representation of the two spatial dimensions of the scene.

Non-Scanning

In non-scanning, a single 2-D sensor output contains all spatial (x,y) and spectral (λ) data. HSI devices for non-scanning yield the full datacube at once, without any scanning. Figuratively speaking, a single snapshot represents a perspective projection of the datacube, from which its three-dimensional structure can be reconstructed. The most prominent benefits of these snapshot hyperspectral imaging systems are the snapshot advantage (higher light throughput) and shorter acquisition time. A number of systems have been designed, including computed tomographic imaging spectrometry (CTIS), fiber-reformatting imaging spectrometry (FRIS), integral field spectroscopy with lenslet arrays (IFS-L), multi-aperture integral field spectrometer (Hyperpixel Array), integral field spectroscopy with image slicing mirrors (IFS-S), image-replicating imaging spectrometry (IRIS), filter stack spectral decomposition (FSSD), coded aperture snapshot spectral imaging (CASSI), image mapping spectrometry (IMS), and multispectral Sagnac interferometry (MSI). However, computational effort and manufacturing costs are high.

Spatiospectral Scanning

In spatiospectral scanning, each 2-D sensor output represents a wavelength-coded ('rainbow-colored', $\lambda = \lambda(y)$), spatial (x,y) map of the scene. A prototype for this technique, introduced in 2014, consists of a camera at some non-zero distance behind a basic slit spectroscope (slit + dispersive element). Advanced spatiospectral scanning systems can be obtained by placing a dispersive element before a spatial scanning system. Scanning can be achieved by moving the whole system relative to the scene, by moving the camera alone, or by moving the slit alone. Spatiospectral scanning unites some advantages of spatial and spectral scanning, thereby alleviating some of their disadvantages.

DISTINGUISHING HYPERSPECTRAL FROM MULTISPECTRAL IMAGING

Hyperspectral imaging is part of a class of techniques commonly referred to as spectral imaging or spectral analysis. Hyperspectral imaging is related to multispectral imaging. The distinction between hyper- and multi-spectral is sometimes based on an arbitrary "number of bands" or on the type of measurement, depending on what is appropriate to the purpose.

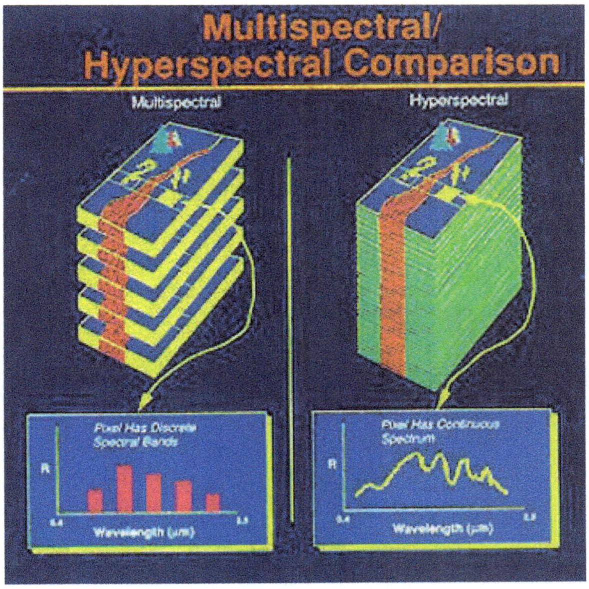

Hyperspectral and Multispectral Differences.

Multispectral imaging deals with several images at discrete and somewhat narrow bands. Being "discrete and somewhat narrow" is what distinguishes multispectral in the visible from color photography. A multispectral sensor may have many bands covering the spectrum from the visible to the longwave infrared. Multispectral images do not produce the "spectrum" of an object. Landsat is an excellent example of multispectral imaging.

Hyperspectral deals with imaging narrow spectral bands over a continuous spectral range, and produce the spectra of all pixels in the scene. So a sensor with only 20 bands can also be hyperspectral when it covers the range from 500 to 700 nm with 20 bands each 10 nm wide. (While a sensor with 20 discrete bands covering the VIS, NIR, SWIR, MWIR, and LWIR would be considered multispectral.)

'Ultraspectral' could be reserved for interferometer type imaging sensors with a very fine spectral resolution. These sensors often have (but not necessarily) a low spatial resolution of several pixels only, a restriction imposed by the high data rate.

Applications

Hyperspectral remote sensing is used in a wide array of applications. Although originally developed for mining and geology (the ability of hyperspectral imaging to identify various minerals makes it ideal for the mining and oil industries, where it can be used to look for ore and oil), it has now spread into fields as widespread as ecology and surveillance, as well as historical manuscript research, such as the imaging of the Archimedes Palimpsest. This technology is continually becoming more available to the public. Organizations such as NASA and the USGS have catalogues of various minerals and their spectral signatures, and have posted them online to make them readily available for researchers.

Agriculture

Although the cost of acquiring hyperspectral images is typically high, for specific crops and in specific climates, hyperspectral remote sensing use is increasing for monitoring the development and health of crops. In Australia, work is under way to use imaging spectrometers to detect grape variety and develop an early warning system for disease outbreaks. Furthermore, work is underway to use hyperspectral data to detect the chemical composition of plants, which can be used to detect the nutrient and water status of wheat in irrigated systems.

Hyperspectral camera embedded on OnyxStar HYDRA-12 UAV from AltiGator

Another application in agriculture is the detection of animal proteins in compound feeds to avoid bovine spongiform encephalopathy (BSE), also known as mad-cow disease. Different studies have been done to propose alternative tools to the reference method of detection, (classical microscopy). One of the first alternatives is near infrared microscopy (NIR), which combines the advantages of microscopy and NIR. In 2004, the first study relating this problem with hyperspectral imaging was published. Hyperspectral libraries that are representative of the diversity of ingredients usually present in the preparation of compound feeds were constructed. These libraries can be used together with chemometric tools to investigate the limit of detection, specificity and reproducibility of the NIR hyperspectral imaging method for the detection and quantification of animal ingredients in feed.

Eye Care

Researchers at the Université de Montréal are working with Photon etc. and Optina Diagnostics to test the use of hyperspectral photography in the diagnosis of retinopathy and macular edema before damage to the eye occurs. The metabolic hyperspectral camera will detect a drop in oxygen consumption in the retina, which indicates potential disease. An ophthalmologist will then be able to treat the retina with injections to prevent any potential damage.

Food Processing

In the food processing industry, hyperspectral imaging, combined with intelligent software, enables digital sorters (also called optical sorters) to identify and remove defects and foreign material (FM) that are invisible to traditional camera and laser sorters. By improving the accuracy of defect and FM removal, the food processor's objective is to enhance product quality and increase yields.

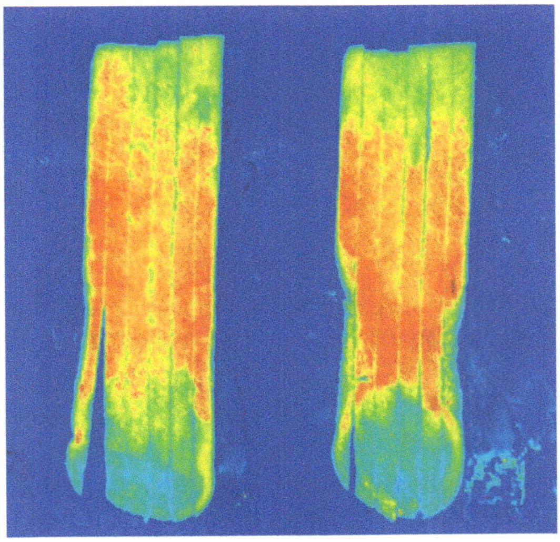

Hyperspectral image of "sugar end" potato strips shows invisible defects

Adopting hyperspectral imaging on digital sorters achieves non-destructive, 100 percent inspection in-line at full production volumes. The sorter's software compares the hyperspectral images collected to user-defined accept/reject thresholds, and the ejection system automatically removes defects and foreign material.

The recent commercial adoption of hyperspectral sensor-based food sorters is most advanced in the nut industry where installed systems maximize the removal of stones, shells and other foreign material (FM) and extraneous vegetable matter (EVM) from walnuts, pecans, almonds, pistachios, peanuts and other nuts. Here, improved product quality, low false reject rates and the ability to handle high incoming defect loads often justify the cost of the technology.

Commercial adoption of hyperspectral sorters is also advancing at a fast pace in the potato processing industry where the technology promises to solve a number of outstanding product quality problems. Work is underway to use hyperspectral imaging to detect "sugar ends," "hollow heart" and "common scab," conditions that plague potato processors.

Mineralogy

Geological samples, such as drill cores, can be rapidly mapped for nearly all minerals of commercial interest with hyperspectral imaging. Fusion of SWIR and LWIR spectral imaging is standard for the detection of minerals in the feldspar, silica, calcite, garnet, and olivine groups, as these minerals have their most distinctive and strongest spectral signature in the LWIR regions.

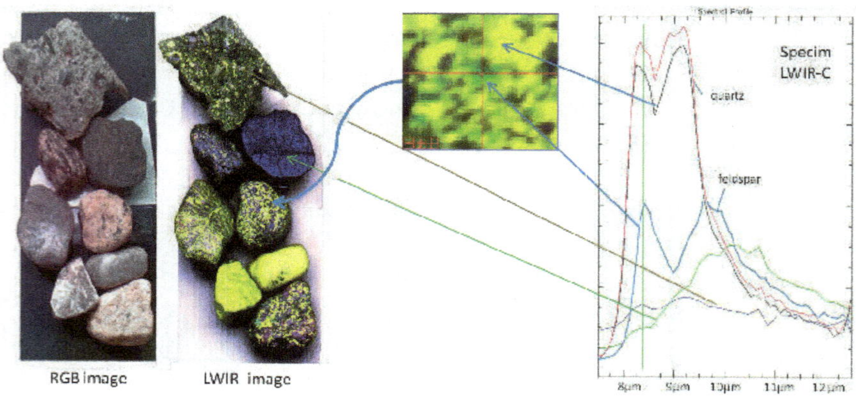

A set of stones is scanned with a Specim LWIR-C imager in the thermal infrared range from 7.7 μm to 12.4 μm. The quartz and feldspar spectra are clearly recognizable.

Hyperspectral remote sensing of minerals is well developed. Many minerals can be identified from airborne images, and their relation to the presence of valuable minerals, such as gold and diamonds, is well understood. Currently, progress is towards understanding the relationship between oil and gas leakages from pipelines and natural wells, and their effects on the vegetation and the spectral signatures. Recent work includes the PhD dissertations of Werff and Noomen.

Surveillance

Hyperspectral surveillance is the implementation of hyperspectral scanning technology for surveillance purposes. Hyperspectral imaging is particularly useful in military surveillance because of countermeasures that military entities now take to avoid airborne surveillance. Aerial surveillance was used by French soldiers using tethered balloons to spy on troop movements during the French Revolutionary Wars, and since that time, soldiers have learned not only to hide from the naked eye, but also to mask their heat signatures to blend into the surroundings and avoid infrared scanning. The idea that drives hyperspectral surveillance is that hyperspectral scanning draws information from such a large portion of the light spectrum that any given object should have a unique spectral signature in at least a few of the many bands that are scanned. The SEALs from NSWDG who killed Osama bin Laden in May 2011 used this technology while conducting the raid (Operation Neptune's Spear) on Osama bin Laden's compound in Abbottabad, Pakistan.

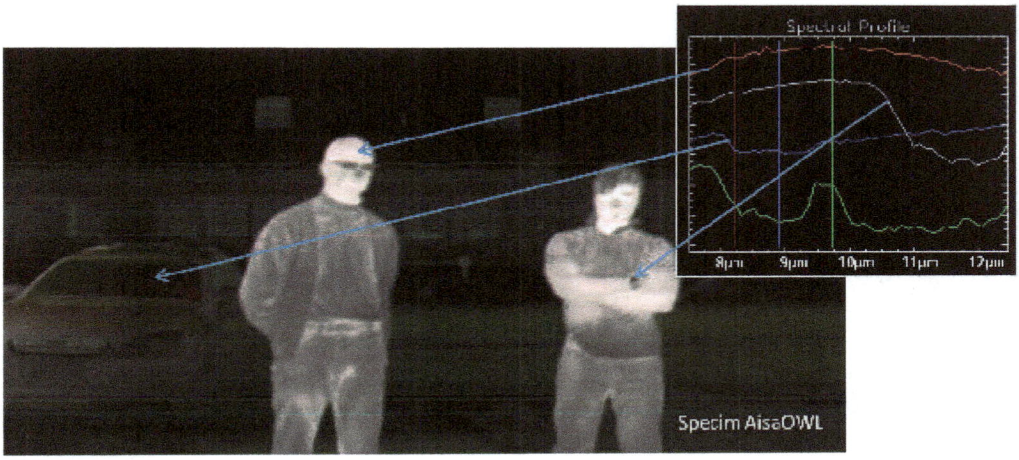

Hyperspectral thermal infrared emission measurement, an outdoor scan in winter conditions, ambient temperature -15°C—relative radiance spectra from various targets in the image are shown with arrows. The infrared spectra of the different objects such as the watch glass have clearly distinctive characteristics. The contrast level indicates the temperature of the object. This image was produced with a Specim LWIR hyperspectral imager.

Traditionally, commercially available thermal infrared hyperspectral imaging systems have needed liquid nitrogen or helium cooling, which has made them impractical for most surveillance applications. In 2010, Specim introduced a thermal infrared hyperspectral camera that can be used for outdoor surveillance and UAV applications without an external light source such as the sun or the moon.

Physics

Physicists use an electron microscopy technique that involves microanalysis using either energy-dispersive X-ray spectroscopy (EDS), electron energy loss spectroscopy (EELS), infrared spectroscopy(IR), Raman spectroscopy, or cathodoluminescence (CL) spectroscopy, in which the entire spectrum measured at each point is recorded. EELS hyperspectral imaging is performed in a scanning transmission electron microscope (STEM); EDS and CL mapping can be performed in STEM as well, or in a scanning electron microscope or electron microprobe (also called an electron probe microanalyzer or EPMA). Often, multiple techniques (EDS, EELS, CL) are used simultaneously.

In a "normal" mapping experiment, an image of the sample is simply the intensity of a particular emission mapped in an XY raster. For example, an EDS map could be made of a steel sample, in which iron X-ray intensity is used for the intensity grayscale of the image. Dark areas in the image would indicate non-iron-bearing impurities. This could potentially give misleading results; if the steel contained tungsten inclusions, for example, the high atomic number of tungsten could result in bremsstrahlung radiation that would make the iron-free areas appear to be rich in iron.

By hyperspectral mapping, instead, the entire spectrum at each mapping point is acquired, and a quantitative analysis can be performed by computer postprocessing of the data, and a quantitative map of iron content produced. This would show which areas contained no iron, despite the anomalous X-ray counts caused by bremsstrahlung. Because EELS core-loss edges are small signals on top of a large background, hyperspectral imaging allows large improvements to the quality of EELS chemical maps.

Similarly, in CL mapping, small shifts in the peak emission energy could be mapped, which would give information regarding slight chemical composition changes or changes in the stress state of a sample.

Astronomy

In astronomy, hyperspectral imaging is used to determine a spatially-resolved spectral image. Since a spectrum is an important diagnostic, having a spectrum for each pixel allows more science cases to be addressed. In astronomy, this technique is commonly referred to as integral field spectroscopy, and examples of this technique include FLAMES and SINFONI on the Very Large Telescope, but also the Advanced CCD Imaging Spectrometer on Chandra X-ray Observatory uses this technique.

Chemical Imaging

Soldiers can be exposed to a wide variety of chemical hazards. These threats are mostly invisible but detectable by hyperspectral imaging technology. The Telops Hyper-Cam, introduced in 2005, has demonstrated this at distances up to 5 km and with concentrations as low as a few ppm.

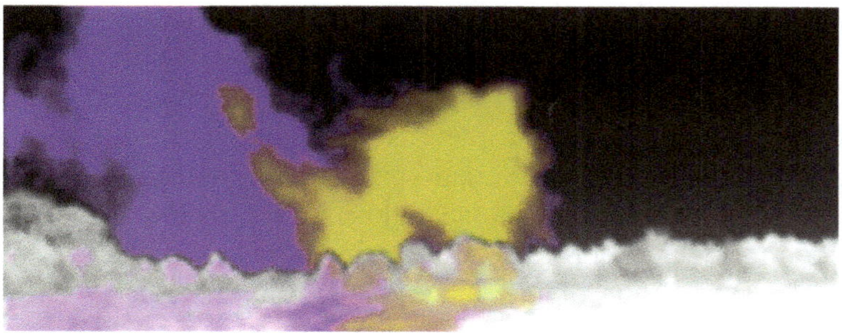

Remote chemical imaging of a simultaneous release of SF_6 and NH_3 at 1.5km using the Telops Hyper-Cam imaging spectrometer

Environment

Most countries require continuous monitoring of emissions produced by coal and oil-fired power plants, municipal and hazardous waste incinerators, cement plants, as well as many other types of industrial sources. This monitoring is usually performed using extractive sampling systems coupled with infrared spectroscopy techniques. Some recent standoff measurements performed allowed the evaluation of the air quality but not many remote independent methods allow for low uncertainty measurements.

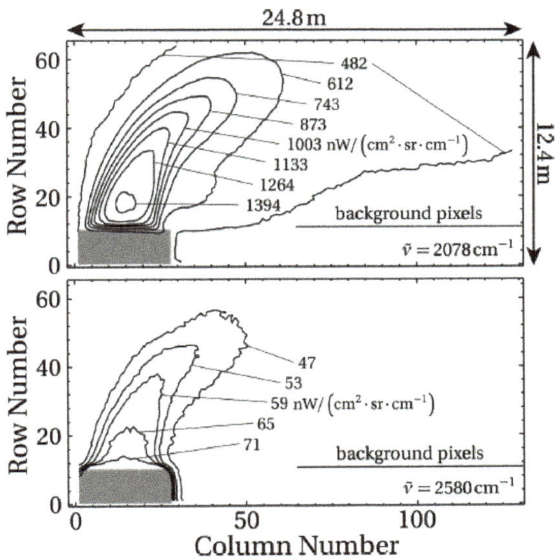

Top panel: Contour map of the time-averaged spectral radiance at 2078 cm^{-1} corresponding to a CO_2 emission line. Bottom panel: Contour map of the spectral radiance at 2580 cm^{-1} corresponding to continuum emission from particulates in the plume. The translucent gray rectangle indicates the position of the stack. The horizontal line at row 12 between columns 64-128 indicate the pixels used to estimate the background spectrum. Measurements made with the Telops Hyper-Cam.

Advantages and Disadvantages

The primary advantage to hyperspectral imaging is that, because an entire spectrum is acquired at each point, the operator needs no prior knowledge of the sample, and postprocessing allows all available information from the dataset to be mined. Hyperspectral imaging can also take advantage of the spatial relationships among the different spectra in a neighborhood, allowing more elaborate spectral-spatial models for a more accurate segmentation and classification of the image.

The primary disadvantages are cost and complexity. Fast computers, sensitive detectors, and large data storage capacities are needed for analyzing hyperspectral data. Significant data storage capacity is necessary since hyperspectral cubes are large, multidimensional datasets, potentially exceeding hundreds of megabytes. All of these factors greatly increase the cost of acquiring and processing hyperspectral data. Also, one of the hurdles researchers have had to face is finding ways to program hyperspectral satellites to sort through data on their own and transmit only the most important images, as both transmission and storage of that much data could prove difficult and costly. As a relatively new analytical technique, the full potential of hyperspectral imaging has not yet been realized.

Software Resources

Open source (in alphabetical order):

- Gerbil, a hyperspectral visualization and analysis framework
- HyperSpy, a Python Hyperspectral Toolbox
- Opticks a remote sensing application.
- R-package Hyperspectral Data Analysis (hsdar)

Commercial (in alphabetical order):

- Erdas Imagine, a remote sensing application for geospatial applications.
- ENVI a remote sensing application.
- FECOM Object Learning Software (OLS), industrial in-line hyperspectral feature processing
- IGiS an image Processing and GIS application
- MicroMSI a remote sensing application.
- A Matlab Hyperspectral Toolbox
- Other Hyperspectral tools in MATLAB
- SOCET GXP

REFERENCES

1. Chein-I Chang (31 July 2003). Hyperspectral Imaging: Techniques for Spectral Detection and Classification. Springer Science & Business Media. ISBN 978-0-306-47483-5.

2. Hans Grahn; Paul Geladi (27 September 2007). Techniques and Applications of Hyperspectral Image Analysis. John Wiley & Sons. ISBN 978-0-470-01087-7.
3. Lu, G; Fei, B (January 2014). "Medical Hyperspectral Imaging: a review" (Online full text article available). Journal of Biomedical Optics. 19 (1): 10901. Bibcode:2014JBO....19a0901L. doi:10.1117/1.JBO.19.1.010901. PMC 3895860Freely accessible. PMID 24441941.
4. "Nikon MicroscopyU – Confocal Microscopy – Spectral Imaging".
5. http://www.bodkindesign.com/wp-content/uploads/2012/09/Hyperspectral-1011.pdf
6. "OSA – Basic slit spectroscope reveals three-dimensional scenes through diagonal slices of hyperspectral cubes".
7. Schurmer, J.H., (Dec 2003), Air Force Research Laboratories Technology Horizons
8. Ellis, J., (Jan 2001) Searching for oil seeps and oil-impacted soil with hyperspectral imagery, Earth Observation Magazine.
9. "SPIE – Journal of Biomedical Optics – Medical hyperspectral imaging: a review".
10. http://www.resonon.com/tutorials.html
11. "HYPERSPECTRAL IMAGING: One-shot camera obtains simultaneous hyperspectral data".
12. "SPIE – Optical Engineering – Snapshot advantage: a review of the light collection improvement for parallel high-dimensional measurement systems".
13. Hyperspectral imaging with spatiospectral images from a simple spectroscope. 12 July 2014 – via YouTube.
14. Smith, R.B. (July 14, 2006), Introduction to hyperspectral imaging with TMIPS, MicroImages Tutorial Web site
15. Lacar, F.M., et al., Use of hyperspectral imagery for mapping grape varieties in the Barossa Valley, South Australia, Geoscience and remote sensing symposium (IGARSS'01) – IEEE 2001 International, vol.6 2875-2877p. doi:10.1109/IGARSS.2001.978191
16. Ferwerda, J.G. (2005), Charting the quality of forage: measuring and mapping the variation of chemical components in foliage with hyperspectral remote sensing, Wageningen University, ITC Dissertation 126, 166p. ISBN 90-8504-209-7
17. Tilling, A.K., et al., (2006) Remote sensing to detect nitrogen and water stress in wheat, The Australian Society of Agronomy
18. Fernández Pierna, J.A., et al., 'Combination of Support Vector Machines (SVM) and Near Infrared (NIR) imaging spectroscopy for the detection of meat and bone meat (MBM) in compound feeds' Journal of Chemometrics 18 (2004) 341-349
19. "HOME".
20. AM Shahidi; et al. (2013). "Regional variation in human retinal vessel oxygen saturation". Elsevier – Experimental Eye Research. 113: 1743–147. doi:10.1016/j.exer.2013.06.001.

21. Higgins, Kevin. "Five New Technologies for Inspection". Food Processing. Retrieved 6 September 2013.
22. Burgstaller, Markus; et al. "Spotlight: Spectral Imaging Sorts 'Sugar-End' Defects". PennWell.
23. Dacal-Nieto, Angel; et al. (2011). Non-Destructive Detection of Hollow Heart in Potatoes Using Hyperspectral Imaging (PDF). pp. 180–187. ISBN 978-3-642-23677-8.
24. Dacal-Nieto, Angel; et al. (2011). Common scab detection on potatoes using an infrared hyperspectral imaging system. pp. 303–312. ISBN 978-3-642-24087-4.
25. Holma, H., (May 2011), Thermische Hyperspektralbildgebung im langwelligen Infrarot Archived July 26, 2011, at the Wayback Machine., Photonik
26. Werff H. (2006), Knowledge based remote sensing of complex objects: recognition of spectral and spatial patterns resulting from natural hydrocarbon seepages, Utrecht University, ITC Dissertation 131, 138p. ISBN 90-6164-238-8
27. Noomen, M.F. (2007), Hyperspectral reflectance of vegetation affected by underground hydrocarbon gas seepage, Enschede, ITC 151p. ISBN 978-90-8504-671-4.
28. "Fleurus (Municipality, Province of Hainaut, Belgium)". CRW Flags Inc.. Retrieved 2010-04-21
29. Marc Ambinder (May 3, 2011). "The secret team that killed bin Laden". National Journal. Retrieved September 12, 2012.
30. Frost&Sullivan, Technical Insights, Aerospace&Defence (Feb 2011): World First Thermal Hyperspectral Camera for Unmanned Aerial Vehicles
31. Specim's Owl sees an invisible object and identifies its materials even in a pitch-dark night.
32. "FLAMES – Fibre Large Array Multi Element Spectrograph". ESO. Retrieved 30 November 2012.
33. "SINFONI – Spectrograph for INtegral Field Observations in the Near Infrared". ESO. Retrieved 30 November 2012.
34. M. Chamberland, V. Farley, A. Vallières, L. Belhumeur, A. Villemaire, J. Giroux et J. Legault, "High-Performance Field-Portable Imaging Radiometric Spectrometer Technology For Hyperspectral imaging Applications," Proc. SPIE 5994, 59940N, September 2005.
35. Farley, V., Chamberland, M., Lagueux, P., et al., "Chemical agent detection and identification with a hyperspectral imaging infrared sensor," Proceedings of SPIE Vol. 6661, 66610L (2007).
36. Kevin C. Gross, Kenneth C Bradley and Glen P. Perram, "Remote identification and quantification of industrial smokestack effluents via imaging Fourier-transform spectroscopy," Environmental Sci Tech, 44, 9390-9397, Oct 2010.
37. A.Picon, O. Ghita, P.F. Whelan, P. Iriondo (2009), Spectral and Spatial Feature Integration for Classification of Non-ferrous Materials in Hyper-spectral

Data,IEEE TRANSACTIONS ON INDUSTRIAL INFORMATICS, VOL. 5, NO. 4, NOVEMBER 2009

38. "Gerbil".
39. "HyperSpy: multi-dimensional data analysis toolbox — HyperSpy".
40. Lehnert, Lukas W.; Meyer, Hanna; Bendix, Joerg (2016-01-12), hsdar: Manage, Analyse and Simulate Hyperspectral Data, retrieved 2016-01-22
41. "FECOM OG – Vision Based Systems".
42. "ISRO participates in "Make In India Week" - ISRO". www.isro.gov.in. Retrieved 2016-09-17.
43. "Matlab Hyperspectral Toolbox".
44. isaacgerg. "Matlab Hyperspectral Toolbox". SourceForge.
45. "Multispectral and Hyperspectral Image Processing with SOCET GXP | GXP". www.geospatialexploitationproducts.com. Retrieved 2016-09-21.

CHAPTER 6

LIDAR Applications

Chapter Contents

LIDAR, which stands for Light Detection and Ranging, is a remote sensing method that uses light in the form of a pulsed laser to measure ranges (variable distances) to the Earth. These light pulses—combined with other data recorded by the airborne system— generate precise, three-dimensional information about the shape of the Earth and its surface characteristics.

A LIDAR instrument principally consists of a laser, a scanner, and a specialized GPS receiver. Airplanes and helicopters are the most commonly used platforms for acquiring LIDAR data over broad areas. Two types of LIDAR are topographic and bathymetric. Topographic LIDAR typically uses a near-infrared laser to map the land, while bathymetric lidar uses water-penetrating green light to also measure seafloor and riverbed elevations.

INTRODUCTION

Lidar (also called LIDAR, LiDAR, and LADAR) is a surveying method that measures distance to a target by illuminating that target with a laser light. The name *lidar*, sometimes considered an acronym of *Light Detection And Ranging* (sometimes *Light Imaging, Detection, And Ranging*), was originally a portmanteau of *light* and *radar*. Lidar is popularly used to make high-resolution maps, with applications in geodesy, geomatics, archaeology, geography, geology, geomorphology, seismology, forestry, atmospheric physics, laser guidance, airborne laser swath mapping (ALSM), and laser altimetry. Lidar sometimes is called *laser scanning* and *3D scanning*, with terrestrial, airborne, and mobile applications.

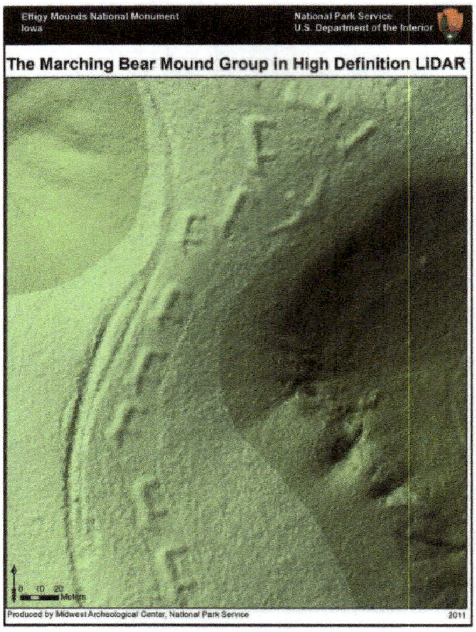

Lidar-derived image of Marching Bears Mound Group, Effigy Mounds National Monument

A FASOR used at the Starfire Optical Range for lidar and laser guide star experiments is tuned to the sodium D2a line and used to excite sodium atoms in the upper atmosphere.

This lidar may be used to scan buildings, rock formations, etc., to produce a 3D model. The lidar can aim its laser beam in a wide range: its head rotates horizontally; a mirror tilts vertically. The laser beam is used to measure the distance to the first object on its path.

HISTORY AND ETYMOLOGY

Lidar originated in the early 1960s, shortly after the invention of the laser, and combined laser-focused imaging with radar's ability to calculate distances by measuring the time for a signal to return. Its first applications came in meteorology, where the National Center for Atmospheric Research used it to measure clouds. The general public became aware of the accuracy and usefulness of lidar systems in 1971 during the Apollo 15 mission, when astronauts used a laser altimeter to map the surface of the moon.

Although some sources treat the word "lidar" as an acronym, the term originated as a portmanteau of "light" and "radar". The first published mention of lidar, in 1963, makes this clear: "Eventually the laser may provide an extremely sensitive detector of particular wavelengths from distant objects. Meanwhile, it is being used to study the moon by 'lidar' (light radar) ..." The *Oxford English Dictionary* supports this etymology.

The interpretation of "lidar" as an acronym ("LIDAR" or "LiDAR") came later, beginning in 1970, based on the assumption that since the base term "radar" originally started as an acronym for "RAdio Detection And Ranging", "LIDAR" must stand for "LIght Detection And Ranging", or for "Laser Imaging, Detection and Ranging". Although the English language no longer treats "radar" as an acronym and printed texts universally present the word uncapitalized, the word "lidar" became capitalized as "LIDAR" or "LiDAR" in some publications beginning in the 1980s. Currently no consensus exists on capitalization, reflecting uncertainty about whether or not "lidar" is an acronym, and if it is an acronym, whether it should appear in lower case, like "radar". Various publications refer to lidar as "LIDAR", "LiDAR", "LIDaR", or "Lidar". The USGS uses both "LIDAR" and "lidar", sometimes in the same document; the *New York Times* uses both "lidar" and "Lidar".

GENERAL DESCRIPTION

Lidar uses ultraviolet, visible, or near infrared light to image objects. It can target a wide range of materials, including non-metallic objects, rocks, rain, chemical compounds, aerosols, clouds and even single molecules. A narrow laser-beam can map physical features with very high resolutions; for example, an aircraft can map terrain at 30-centimetre (12 in) resolution or better.

Lidar has been used extensively for atmospheric research and meteorology. Lidar instruments fitted to aircraft and satellites carry out surveying and mapping – a recent example being the U.S. Geological Survey Experimental Advanced Airborne Research Lidar. NASA has identified lidar as a key technology for enabling autonomous precision safe landing of future robotic and crewed lunar-landing vehicles.

Wavelengths vary to suit the target: from about 10 micrometers to the UV (approximately 250 nm). Typically light is reflected via backscattering. Different types of scattering are used for different lidar applications: most commonly Rayleigh scattering, Mie scattering, Raman scattering, and fluorescence. Based on different kinds of backscattering, the lidar can be accordingly called Rayleigh Lidar, Mie Lidar, Raman Lidar, Na/Fe/K Fluorescence Lidar, and so on. Suitable combinations of wavelengths can allow for remote mapping of atmospheric contents by identifying wavelength-dependent changes in the intensity of the returned signal.

Design

In general there are two kinds of lidar detection schemes: "incoherent" or direct energy detection (which is principally an amplitude measurement) and coherent detection (which is best for Doppler, or phase sensitive measurements). Coherent systems generally use optical heterodyne detection, which, being more sensitive than direct detection, allows them to operate at a much lower power but at the expense of more complex transceiver requirements.

In both coherent and incoherent lidar, there are two types of pulse models: *micropulse lidar* systems and *high energy* systems. Micropulse systems have developed as a result of the ever-increasing amount of computer power available combined with advances in laser technology. They use considerably less energy in the laser, typically on the order of one microjoule, and are often "eye-safe," meaning they can be used without safety precautions. High-power systems are common in atmospheric research, where they are widely used for measuring many atmospheric parameters: the height, layering and densities of clouds, cloud particle properties (extinction coefficient, backscatter coefficient, depolarization), temperature, pressure, wind, humidity, trace gas concentration (ozone, methane, nitrous oxide, etc.).

There are several major components to a lidar system:

- Laser — 600–1000 nm lasers are most common for non-scientific applications. They are inexpensive, but since they can be focused and easily absorbed by the eye, the maximum power is limited by the need to make them eye-safe. Eye-safety is often a requirement for most applications. A common alternative, 1550 nm lasers, are eye-safe at much higher power levels since this wavelength is not focused by the eye, but the detector technology is less advanced and so these wavelengths are generally used at longer ranges and lower accuracies. They are also used for military applications as 1550 nm is not visible in night vision goggles, unlike the shorter 1000 nm infrared laser. Airborne topographic mapping lidars generally use 1064 nm diode pumped YAG lasers, while bathymetric systems generally use 532 nm frequency doubled diode pumped YAG lasers because 532 nm penetrates water with much less attenuation than does 1064 nm. Laser settings include the laser repetition rate (which controls the data collection speed). Pulse length is generally an attribute of the laser cavity length, the number of passes required through the gain material (YAG, YLF, etc.), and Q-switch speed. Better target resolution is achieved with shorter pulses, provided the lidar receiver detectors and electronics have sufficient bandwidth.
- Scanner and optics — How fast images can be developed is also affected by the speed at which they are scanned. There are several options to scan the azimuth and elevation, including dual oscillating plane mirrors, a combination with a polygon mirror, a dual axis scanner. Optic choices affect the angular resolution and range that can be detected. A hole mirror or a beam splitter are options to collect a return signal.
- Photodetector and receiver electronics — Two main photodetector technologies are used in lidars: solid state photodetectors, such as silicon avalanche photodiodes, or photomultipliers. The sensitivity of the receiver is another parameter that has to be balanced in a lidar design.
- Position and navigation systems — Lidar sensors that are mounted on mobile platforms such as airplanes or satellites require instrumentation to determine the absolute position and orientation of the sensor. Such devices generally include a Global Positioning System receiver and an Inertial Measurement Unit (IMU).

3D imaging can be achieved using both scanning and non-scanning systems. "3D gated viewing laser radar" is a non-scanning laser ranging system that applies a pulsed laser and a fast gated camera. Research has begun for virtual beam steering using DLP technology.

Imaging lidar can also be performed using arrays of high speed detectors and modulation sensitive detector arrays typically built on single chips using CMOS and hybrid CMOS/CCD fabrication techniques. In these devices each pixel performs some local processing such as demodulation or gating at high speed, downconverting the signals to video rate so that the array may be read like a camera. Using this technique many thousands of pixels / channels may be acquired simultaneously. High resolution 3D lidar cameras use homodyne detection with an electronic CCD or CMOS shutter.

A coherent imaging lidar uses synthetic array heterodyne detection to enable a staring single element receiver to act as though it were an imaging array.

In 2014 Lincoln Laboratory announced a new imaging chip with more than 16,384 pixels, each able to image a single photon, enabling them to capture a wide area in a single image. An earlier generation of the technology with one-quarter as many pixels was dispatched by the U.S. military after the January 2010 Haiti earthquake; a single pass by a business jet at 3,000 meters (10,000 ft.) over Port-au-Prince was able to capture instantaneous snapshots of 600-meter squares of the city at 30 centimetres (12 in), displaying the precise height of rubble strewn in city streets. The new system is another 10x faster. The chip uses indium gallium arsenide (InGaAs), which operates in the infrared spectrum at a relatively long wavelength that allows for higher power and longer ranges. In many applications, such as self-driving cars, the new system will lower costs by not requiring a mechanical component to aim the chip. InGaAs uses less hazardous wavelengths than conventional silicon detectors, which operate at visual wavelengths.

TYPES OF APPLICATIONS

Lidar has a wide range of applications which can be divided into airborne and terrestrial types. These different types of applications require scanners with varying specifications based on the data's purpose, the size of the area to be captured, the range of measurement desired, the cost of equipment, and more.

Airborne Lidar

Airborne lidar (also *airborne laser scanning*) is when a laser scanner, while attached to a plane during flight, creates a 3D point cloud model of the landscape. This is currently the most detailed and accurate method of creating digital elevation models, replacing photogrammetry. One major advantage in comparison with photogrammetry is the ability to filter out reflections from vegetation from the point cloud model to create a digital surface model which represents ground surfaces such as rivers, paths, cultural heritage sites, etc., which are concealed by trees. Within the category of airborne lidar, there is sometimes a distinction made between high-altitude and low-altitude applications, but the main difference is a reduction in both accuracy and point density of data acquired at higher altitudes. Airborne lidar can also be used to create bathymetric models in shallow water.

The main constituents of airborne lidar include digital elevation models (DEM) and digital survey models (DSM). The points and ground points are the vectors of discrete points while DEM and DSM are interpolated raster grids of discrete points. The process also involves capturing of digital aerial photographs. In order to interpret deep seated landslides for example, under the cover of vegetation, scarps, tension cracks or tipped trees air borne lidar is used. Air borne lidar digital elevation models can see through the canopy of forest cover, perform detailed measurements of scarps, erosion and tilting of electric poles.

Airborne lidar data is processed using a toolbox(TIFFS-Toolbox for Lidar Data Filtering and Forest Studies) for lidar data filtering and terrain study software. The data is interpolated to digital terrain models using the software. The laser is directed at the region to be mapped and each point's height above the ground is calculated by subtracting the original z-coordinate from the corresponding digital terrain model elevation. Based on this height above the ground the non-vegetation data is obtained which may include objects such as buildings, electric power lines, flying birds etc. The rest of the points are treated as vegetation and used for modeling and mapping. Within each of these plots, lidar metrics are calculated by calculating statistics such as mean, standard deviation, skewness, percentiles, quadratic mean etc.

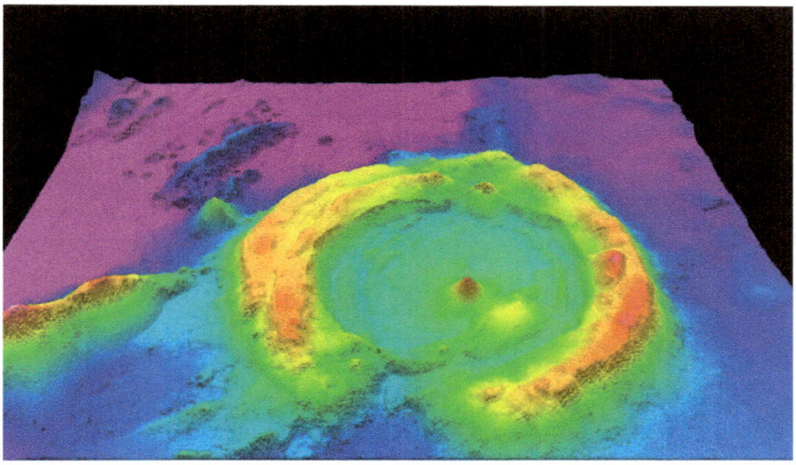

Airborne Lidar Bathymetric Technology-High-resolution multibeam lidar map showing spectacularly faulted and deformed seafloor geology, in shaded relief and coloured by depth.

Airborne Lidar Bathymetric Technology

The airborne lidar bathymetric technological system involves the measurement of time of flight of a signal from a source to its return to the sensor. The data acquisition technique involves a sea floor mapping component and a ground truth component that includes video transects and sampling. It works using a near infrared laser beam. Two beams are projected onto a fast rotating mirror, which creates an array of points. One of the beams penetrates the water and also detects the bottom surface of the water under favorable conditions.

The data obtained shows the full extent of the land surface exposed above the sea floor. This technique is extremely useful as it will play an important role in the major sea floor mapping program. The mapping yields onshore topography as well as under water elevations. Sea floor reflectance imaging is another solution product from this system which can benefit mapping of underwater habitats. This technique has been used for three dimensional image mapping of California's waters using a hydrographic lidar.

LIDAR scanning performed with the OnyxStar FOX-C8 HD UAV from AltiGator

Drones are now being used with laser scanners, as well as other remote sensors, as a more economical method to scan smaller areas. The possibility of drone remote sensing also eliminates any danger that crews of a manned aircraft may be subjected to in difficult terrain or remote areas.

Terrestrial Lidar

Terrestrial applications of lidar (also *terrestrial laser scanning*) happen on the Earth's surface and can be both stationary or mobile. Stationary terrestrial scanning is most common as a survey method, for example in conventional topography, monitoring, cultural heritage documentation and forensics. The 3D point clouds acquired from these types of scanners can be matched with digital images taken of the scanned area from the scanner's location to create realistic looking 3D models in a relatively short time when compared to other technologies. Each point in the point cloud is given the colour of the pixel from the image taken located at the same angle as the laser beam that created the point.

Mobile lidar (also *mobile laser scanning*) is when two or more scanners are attached to a moving vehicle to collect data along a path. These scanners are almost always paired with other kinds of equipment, including GNSS receivers and IMUs. One example application is surveying streets, where power lines, exact bridge heights, bordering trees, etc. all need to be taken into account. Instead of collecting each of these measurements individually in the field with a tachymeter, a 3D model from a point cloud can be created where all of the measurements needed can be made, depending on the quality of the data collected. This eliminates the problem of forgetting to take a measurement, so long as the model is available, reliable and has an appropriate level of accuracy.

Terrestrial lidar mapping involves a process of occupancy grid map generation. The process involves an array of cells divided into grids which employs a process to store the height values when lidar data falls into the respective grid cell. A binary map is then created by applying a particular threshold to the cell values for further processing. The next step is to process the radial distance and z-coordinates from each scan to identify which 3D points correspond to each of the specified grid cell leading to the process of data formation.

Data Formation

For the autonomous applications a lidar with a 3D scanner is used for obstacle detection and collision avoidance. The scanner measures radial distance at different angular resolutions. The lidar emits a single laser ray and uses an interior rotating mirror to distribute the laser ray covering a large field of view.

Object Detection

The object detection procedure consists of three steps: Laser point feature calculation and prior filtering, 3D segmentation, object classification and 2D position calculation. The features for segmentation are calculated in the vicinity of a fixed radial distance in 2D and 3D. These include:

Amplitude Density

This involves the percentage of points in the vicinity with amplitude values below a particular threshold. Points with a density lower than the threshold are removed from segmentation and further processing.

Height above Local Minimum in 2D Search Radius

In this case the points which have a height lower than a certain threshold are removed. On segmentation of the point cloud, all the segments satisfying the object based criteria are included in individual objects.

Advantages

The main advantages of object detection and data formation are that high point density and high spatial resolution is obtained. Also the accuracy of the acquired data sets is increased to a considerable extent.

Disadvantages

Although significant amount of research has been done for object classification from 3D point clouds, the direct extraction of individual points from Lidar inputs has not been achieved.

Object Detection for Transportation System

In transportation systems, to ensure vehicle and passenger safety and to develop electronic systems that deliver driver assistance, understanding vehicle and surrounding environment is essential. LIDAR systems plays an important role in safety of transportation systems. Lots of electronic systems which add to the driver assistance and vehicle safety such as Adaptive Cruise Control (ACC), Emergency Brake Assist, Anti-lock Braking System (ABS) depends on the detection of vehicle environment to act autonomously or semi-autonomously. LIDAR mapping and estimation achieve this.

Basics overview: Current LIDAR systems use rotating hexagonal mirrors which split the laser beam. The upper three beams are used for vehicle and obstacles ahead and the lower beams are used to detect lane markings and road features. The major advantage of using LIDAR is that the spatial structure is obtained and this data can be fused with other sensors like RADAR etc. to get a better picture of the vehicle environment in terms of static and dynamic properties of the objects present in the environment.

Applications

There are a wide variety of applications for lidar, in addition to the applications listed below, as it is often mentioned in National lidar dataset programs.

This lidar-equipped mobile robot uses its lidar to construct a map and avoid obstacles.

Agriculture

Lidar also can be used to help farmers determine which areas of their fields to apply costly fertilizer. Lidar can create a topographical map of the fields and reveals the slopes and sun exposure of the farm land. Researchers at the Agricultural Research Service blended this topographical information with the farmland yield results from previous years. From this information, researchers categorized the farm land into high-, medium-, or low-yield zones. This technology is valuable to farmers because it indicates which areas to apply the expensive fertilizers to achieve the highest crop yield.

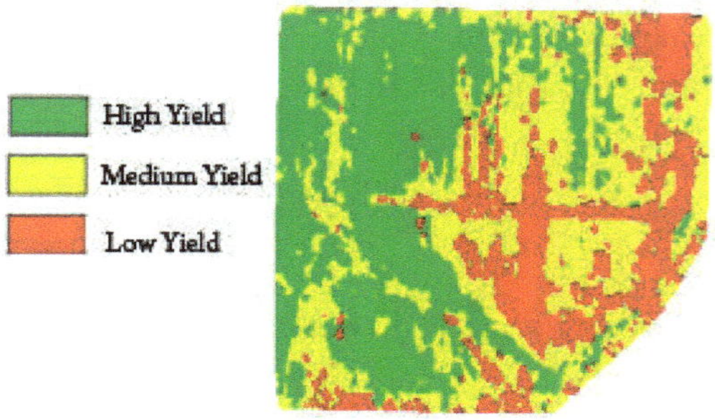

Agricultural Research Service scientists have developed a way to incorporate lidar with yield rates on agricultural fields. This technology will help farmers improve their yields by directing their resources toward the high-yield sections of their land.

Another application of lidar beyond crop health and terrain mapping is crop mapping in orchards and vineyards. Vehicles equipped with lidar sensors can detect foliage growth to determine if pruning or other maintenance needs to take place, detect variations in fruit production, or perform automated tree counts.

Lidar is useful in GPS-denied situations, such as in nut and fruit orchards where GPS signals to farm equipment featuring precision agriculture technology or a driverless tractor may be partially or completely blocked by overhanging foliage. Lidar sensors can detect the edges of rows so that farming equipment can continue moving until GPS signal can be reestablished.

Plant species Classification Using a 3D Lidar Sensor and Machine Learning.

This is a methodology for differentiating plant species using a 3D lidar sensor and machine learning. The procedure involves the application of logistic regression functions, support vector machines and neural networks.

Problem Statement

Autonomous robots have been used for a variety of purposes in agriculture ranging from seed and fertilizer dispersions, sensing techniques as well as crop scouting for the task of weed control. In order to execute weed control one of the major factors is the detection and classification of the plant and species. In addition it is difficult to differentiate the characteristics and appearances of the plant and convert the data into computer understandable form. Also, it is not possible to accurately detect structures of the plant in 3D data. The problem is solved by classifying the plant species by using a set of example plants and machine learning methods.

Algorithm Overview

The input to the system are 3D point clouds with range and reflectance values from a 3D laser sensor. The species of the plant is known by applying preprocessing techniques to provide the point cloud for further processing. If the species is already known then the extracted features are added as new data. The particular plant species is labeled and its features are initially stored as an example to identify the particular plant species in the real environment.

Pre-Processing

In this process, there are three steps. The first step is to identify the ground plane and the points returning from the ground plane are eliminated. In the second step the remaining points are grouped as one and the individual points are detected. The third step involves the identification of the plant species. An individual coordinate system is developed for each plant. The x and y coordinates are placed on the ground while the z-axis is along the center of the plant. In this way individual plants in row cultivations are separated. The next step is to identify the plant space.

Feature Extraction

To understand a set of features to identify the appearance of the plant species the criteria of size and rotation invariance of the 3D laser sensor is used. The set of features are divided into two groups: Reflectance and geometrical features

Reflectance Features

Plant species have different variations based on their reflectance values. It is an important tool for plant classification. The reflectance value encodes information about the leaf size and shape. For each of the measuring points a reflectance value between 0 and 1 are determined.

Geometrical Features

The laser sensor also provides a range of data which can be transformed into 3D coordinates. Using the coordinates, the shape, dimension and structure of the plant can be characterized.

Classification

Classification of the plants is based on machine learning. For the learning process a training set is generated with known samples. Machine learning toolbox WEKA is used for this purpose.

Tested plants

The training set is generated with the help of potted plants available in common nurseries. Different plant species is chosen and the corresponding plant cloud is developed. Besides the plant the ground coordinate system is also sensed to determine the plant coordinate system that is essential for feature calculation.

Advantages

This method for distinguishing plant species is highly efficient since it uses a low resolution 3D Lidar sensor and supervised learning. It includes easy to compute feature set with common statistical features which are independent of the plant size.

Archaeology

Lidar has many applications in the field of archaeology including aiding in the planning of field campaigns, mapping features beneath forest canopy, and providing an overview of broad, continuous features that may be indistinguishable on the ground. Lidar can also provide archaeologists with the ability to create high-resolution digital elevation models (DEMs) of archaeological sites that can reveal micro-topography that are otherwise hidden by vegetation. Lidar-derived products can be easily integrated into a Geographic Information System (GIS) for analysis and interpretation. For example, at Fort Beauséjour - Fort Cumberland National Historic Site, Canada, previously undiscovered archaeological features below forest canopy have been mapped that are related to the siege of the Fort in 1755. Features that could not be distinguished on the ground or through aerial photography were identified by overlaying hillshades of the DEM created with artificial illumination from various angles. With lidar, the ability to produce high-resolution datasets quickly and relatively cheaply can be an advantage. Beyond efficiency, its ability to penetrate forest canopy has led to the discovery of features that were not distinguishable through traditional geo-spatial methods and are difficult to reach through field surveys, as in work at Caracol by Arlen Chase and his wife Diane Zaino Chase. The intensity of the returned signal can be used to detect features buried under flat vegetated surfaces such as fields, especially when mapping using the infrared spectrum. The presence of these features affects plant growth and thus the amount of infrared light reflected back. In 2012, lidar was used by a team attempting to find the legendary city of La Ciudad Blanca in the Honduran jungle. During a seven-day mapping period, they found evidence of extensive man-made structures that had eluded ground searches for hundreds of years. In June 2013 the rediscovery of the city of Mahendraparvata was announced. In another study, lidar was used to reveal stone walls, building foundations, abandoned roads, and other features of the landscape in southern New England, USA that had been obscured in aerial photography by the region's dense forest canopy. In May 2012, lidar was used to locate a previously unknown ruined city in the La Mosquitia region of Honduras. In Cambodia Lidar data were used by Demian Evans and Roland Fletcher to reveal anthropogenic changes to Angkor landscape

Autonomous Vehicles

Autonomous vehicles use lidar for obstacle detection and avoidance to navigate safely through environments, using rotating laser beams. Cost map or point cloud outputs from the lidar sensor provide the necessary data for robot software to determine where potential obstacles exist in the environment and where the robot is in relation to those potential obstacles.

Forecast 3D Laser System uses a SICK LMC lidar sensor

Singapore's *Singapore-MIT Alliance for Research and Technology (SMART)* is actively developing technologies for autonomous lidar vehicles. Examples of companies that produce lidar sensors commonly used in robotics or vehicle automation are Sick and Hokuyo. Examples of obstacle detection and avoidance products that leverage lidar sensors are the Autonomous Solution, Inc. Forecast 3D Laser System and Velodyne HDL-64E.

The very first generations of automotive adaptive cruise control systems used only lidar sensors.

Approaches of Processing LIDAR DATA

Below mentioned are various approaches of processing LIDAR data and utilizing it along with data from other sensors through sensor fusion to detect the vehicle environment conditions.

LiDAR image shows road contour, elevation and roadside vegetation.

GRID Based Processing using 3D LIDAR and Fusion with RADAR Measurement

In this method, proposed by Philipp Lindner and Gerd Wanielik, laser data is processed using a multidimensional occupancy grid. Data from a 4 layer laser is pre-processed on signal level and then processed on higher level to extract the features of the obstacles. A combination of 2 and 3-dimensional grid structure is utilized and the space in these structures are tessellated into several discrete cells. This way huge amount of raw measurement data can be handled effectively by collecting them in spatial containers, which are the cells of the evidence grid. Each cell is associated with a probability measure that identifies the cell occupation. This probability is calculated by using the range measurement of lidar sensor obtained over time and a new range measurement, that is using Bayes' theorem the latest and previous range measurements of lidar sensor are related. A two dimensional grid can observe an obstacle in front of it, but cannot observe the space behind the obstacle. That's why state of occupancy behind the obstacle is, hence the unknown state is assigned a probability of 0.5. By introducing the third dimension or in other terms using a multi-layer laser, the spatial configuration of an object could be mapped into the grid structure up to a complexity. This is achieved by transferring the measurement points into a 3 dimensional grid. The grid cells which are occupied will possess a probability greater than 0.5 and the mapping would be color coded based on the probability. The cells which are not occupied will possess a probability less than 0.5 and this area will usually be white space. This measurement is then transformed to grid coordinate system by using the sensor position on the vehicle and the vehicle position in the world coordinate system. The coordinates of the sensor depends upon its location on the vehicle and the coordinates of the vehicle is computed using egomotion estimation, which is estimating the vehicle motion relative to rigid scene. But for this method the grid profile must be defined. The grid cells touched by the laser beam when transmitted are calculated by applying Bresenham's line algorithm. To obtain the spatial extended structure a connected component analysis of these cells is performed. This information is then passed on to a rotating caliper algorithm to obtain the spatial characteristics of the object. In addition to the LIDAR detection, RADAR data obtained by using two short range radars is integrated to get the dynamic properties such as velocity of the object. The measurements are assigned to the object using a potential distance function.

Advantages and Disadvantages

The geometric features of the objects are extracted efficiently, from the measurements obtained by the 3D occupancy grid, using rotating caliper algorithm. Fusing the radar data to the lidar measurements give information about the dynamic properties of the obstacle such as velocity and location of the obstacle with respect to the sensor location which helps the vehicle or the driver decide the action to be performed in order to ensure safety. The only concern is the computational requirement to implement this data processing technique. It can be implemented in real time and is been proven efficient if the 3D occupancy grid size is considerable restricted. But this can be improved to an even wider range by using dedicated spatial datastructures that manipulate the spatial data more effectively, for the 3D grid representation.

Fusion of 3D LIDAR and Color Camera for Multiple Object Detection and Tracking

The framework proposed in this method by Soonmin Hwang et al., is split into four steps. First, the data from camera and 3D LIDAR is given as input to the system. Both inputs from LIDAR and camera are parallelly obtained and the color image from the camera is calibrated with the LIDAR. To improve the efficiency, horizontal 3D point sampling is applied as pre-processing. Second, the segmentation stage is where the entire 3D points are divided into several groups per the distance from the sensor and local planes from close plane to far plane are sequentially estimated. The local planes are estimated using statistical analysis. The group of points closer to the sensor are used to compute the initial plane. By using the current local plane, the next local plane is estimated by iterative update. The object proposals in the 2D image are used to separate foreground objects from background. For faster and accurate detection and tracking Binarized Normed Gradients for Objectness Estimation at 300fps is used. BING is a combination of normed gradient and its binarized version which speeds up the feature extraction and testing process, to estimate the objectness of an image window. This way the foreground and background objects are separated. To form objects after estimating the objectness of an image using BING, the 3D points are grouped or clustered. Clustering is done using DBSCAN (Density-Based Spatial Clustering of Applications with Noise) algorithm which could be robust due to its less-parametric characteristic. Using the clustered 3D points, i.e. 3D segment, more accurate region-of-interests (ROIs) are generated by projecting 3D points on the 2D image. The third step is detection, which is broadly divided into two parts. First is object detection in 2D image which is achieved using Fast R-CNN as this method doesn't need training and it also considers an image and several regions of interest. Second is object detection in 3D space which is done by using the spin image method. This method extracts local and global histograms to represent a certain object. To merge the results of 2D image and 3D space object detection, same 3D region is considered and two independent classifiers from 2D image and 3D space are applied to the considered region. Scores calibration is done to get a single confidence score from both detectors. This single score is obtained in the form of probability. The final step is tracking. This is done by associating moving objects in present and past frame. For object tracking, segment matching is adopted. Features such as mean, standard deviation, quantized color histograms, volume size and number of 3D points of a segment are computed. Euclidean distance is used to measure differences between segments. To judge the appearance and disappearance of an object, similar segments (obtained based on the Euclidean distance) from two different frames are taken and the physical distance and dissimilarity scores are calculated. If the scores go beyond a range for every segment in previous frame, the object being tracked is considered to have disappeared.

Advantages and Disadvantages

The advantages of this method are Using 2D image and 3D data together, F 1-score (which gives a measure of test's accuracy), average precision (AP) are higher than that when only 3D data from LIDAR is used. These scores are conventional measurements which judge the framework. The drawback of this method is the usage of BING for object proposal estimation as BING predicts a small set of object bounding boxes.

Obstacle Detection and Road Environment Recognition using LIDAR

This method proposed by Kun Zhou et al. not only focuses on object detection and tracking but also recognizes lane marking and road features. As mentioned earlier the LIDAR systems use rotating hexagonal mirrors which split the laser beam into 6 beams. The upper 3 layers are used to detect the forward objects such as vehicles and roadside objects. The sensor is made of weather-robust head. The data detected by LIDAR are clustered to several segments and tracked by Kalman filter. Data clustering here is done based on characteristics of each segment based on object model, which distinguish different objects such as vehicles, signboards etc. These characteristics include the dimensions of the object etc. The reflectors on the rear edges of vehicles are used to differentiate vehicles from other objects. Object tracking is done using a 2-stage Kalman filter considering the stability of tracking and the accelerated motion of objects LIDAR reflective intensity data is also used for curb detection by making use of robust regression to deal with occlusions The road marking is detected using a modified Otsu method by distinguishing rough and shiny surfaces.

Advantages

The advantages of this method Roadside reflectors which indicate lane border are sometimes hidden due to various reasons. Therefore, other information is needed to recognize the road border. The LIDAR used in this method can measure the reflectivity from the object. Hence, with this data road border can also be recognized. Also the usage of sensor with weather-robust head helps detecting the objects even in bad weather conditions.

Advantages of using LIDAR Measurement

LIDAR measurements help identify the spatial structure of the obstacle. This helps distinguish objects based on size and estimate the impact of driving over it.

LIDAR system provides better range and a large field of view which helps detecting obstacles on the curves. This is one major advantage over RADAR systems which has a narrow field of view. Fusion of LIDAR measurement with different sensors makes the system robust and useful in real-time applications as LIDAR dependent system will not be able to estimate the dynamic information about the detected object.

It has been shown that lidar can be manipulated, such that self-driving cars are tricked into taking evasive action.

Biology and Conservation

Lidar has also found many applications in forestry. Canopy heights, biomass measurements, and leaf area can all be studied using airborne lidar systems. Similarly, lidar is also used by many industries, including Energy and Railroad, and the Department of Transportation as a faster way of surveying. Topographic maps can also be generated readily from lidar, including for recreational use such as in the production of orienteering maps.

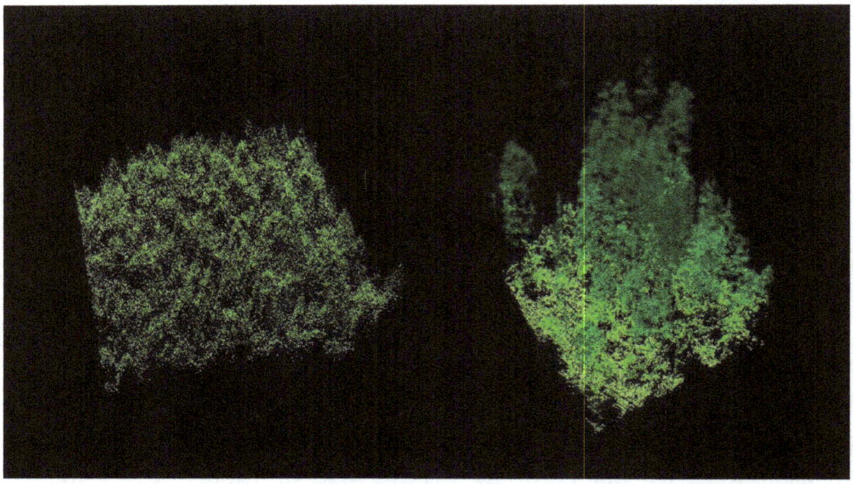

Lidar imaging comparing old-growth forest (right) to a new plantation of trees (left).

In addition, the Save-the-Redwoods League is undertaking a project to map the tall redwoods on the Northern California coast. Lidar allows research scientists to not only measure the height of previously unmapped trees but to determine the biodiversity of the redwood forest. Stephen Sillett, who is working with the League on the North Coast lidar project, claims this technology will be useful in directing future efforts to preserve and protect ancient redwood trees.

Geology and Soil Science

High-resolution digital elevation maps generated by airborne and stationary lidar have led to significant advances in geomorphology (the branch of geoscience concerned with the origin and evolution of the Earth surface topography). The lidar abilities to detect subtle topographic features such as river terraces and river channel banks, to measure the land-surface elevation beneath the vegetation canopy, to better resolve spatial derivatives of elevation, and to detect elevation changes between repeat surveys have enabled many novel studies of the physical and chemical processes that shape landscapes. In 2005 the Tour Ronde in the Mont Blanc massif became the first high alpine mountain on which lidar was employed to monitor the increasing occurrence of severe rock-fall over large rock faces allegedly caused by climate change and degradation of permafrost at high altitude.

In geophysics and tectonics, a combination of aircraft-based lidar and GPS has evolved into an important tool for detecting faults and for measuring uplift. The output of the two technologies can produce extremely accurate elevation models for terrain - models that can even measure ground elevation through trees. This combination was used most famously to find the location of the Seattle Fault in Washington, United States. This combination also measures uplift at Mt. St. Helens by using data from before and after the 2004 uplift. Airborne lidar systems monitor glaciers and have the ability to detect subtle amounts of growth or decline. A satellite-based system, the NASA ICESat, includes a lidar sub-system for this purpose. The NASA Airborne Topographic Mapper is also used extensively to monitor glaciers and perform coastal change analysis. The combination is also used by soil scientists while creating a soil survey. The detailed terrain modeling allows soil scientists to see slope changes and landform breaks which indicate patterns in soil spatial relationships.

Atmospheric Remote Sensing and Meteorology

Initially based on ruby lasers, lidar for meteorological applications was constructed shortly after the invention of the laser and represent one of the first applications of laser technology. Lidar technology has since expanded vastly in capability and lidar systems are used to perform a range of measurements that include profiling clouds, measuring winds, studying aerosols and quantifying various atmospheric components. Atmospheric components can in turn provide useful information including surface pressure (by measuring the absorption of oxygen or nitrogen), greenhouse gas emissions (carbon dioxide and methane), photosynthesis (carbon dioxide), fires (carbon monoxide) and humidity (water vapor). Atmospheric lidars can be either ground-based, airborne or satellite depending on the type of measurement.

Atmospheric lidar remote sensing works in two ways -

- by measuring backscatter from the atmosphere, and
- by measuring the scattered reflection off the ground (when the lidar is airborne) or other hard surface.

Backscatter from the atmosphere directly gives a measure of clouds and aerosols. Other derived measurements from backscatter such as winds or cirrus ice crystals require careful selecting of the wavelength and/or polarization detected. *Doppler Lidar* and *Rayleigh Doppler Lidar* are used to measure temperature and/or wind speed along the beam by measuring the frequency of the backscattered light. The Doppler broadening of gases in motion allows the determination of properties via the resulting frequency shift. Scanning lidars, such as the conical-scanning NASA HARLIE LIDAR, have been used to measure atmospheric wind velocity. The ESA wind mission ADM-Aeolus will be equipped with a Doppler lidar system in order to provide global measurements of vertical wind profiles. A doppler lidar system was used in the 2008 Summer Olympics to measure wind fields during the yacht competition.

Doppler lidar systems are also now beginning to be successfully applied in the renewable energy sector to acquire wind speed, turbulence, wind veer and wind shear data. Both pulsed and continuous wave systems are being used. Pulsed systems use signal timing to obtain vertical distance resolution, whereas continuous wave systems rely on detector focusing.

The term *eolics* has been proposed to describe the collaborative and interdisciplinary study of wind using computational fluid mechanics simulations and Doppler lidar measurements.

The ground reflection of an airborne lidar gives a measure of surface reflectivity (assuming the atmospheric transmittance is well known) at the lidar wavelength. However, the ground reflection is typically used for making absorption measurements of the atmosphere. "Differential absorption lidar" (DIAL) measurements utilize two or more closely spaced (<1 nm) wavelengths to factor out surface reflectivity as well as other transmission losses, since these factors are relatively insensitive to wavelength. When tuned to the appropriate absorption lines of a particular gas, DIAL measurements can be used to determine the concentration (mixing ratio) of that particular gas in the atmosphere. This is referred to as an *Integrated Path Differential Absorption* (IPDA) approach, since it is a measure of the integrated absorption along the entire lidar path. IPDA lidars can be either pulsed or CW and typically use two or more wavelengths. IPDA lidars have been used for remote sensing of carbon dioxide and methane.

Synthetic array lidar allows imaging lidar without the need for an array detector. It can be used for imaging Doppler velocimetry, ultra-fast frame rate (MHz) imaging, as well as for speckle reduction in coherent lidar. An extensive lidar bibliography for atmospheric and hydrospheric applications is given by Grant.

Scheimpflug Principles

Another lidar technique for atmospheric remote sensing has emerged. It is based on Scheimpflug principles referred to as scheimpflug lidar (slidar).

"*The implication of the Scheimpflug principle is that when a laser beam is transmitted into the atmosphere, the backscattering echo of the entire illuminating probe volume is still in focus simultaneously without diminishing the aperture as long as the object plane, image plane and the lens plane intersect with each other*". A two dimensional CCD/CMOS camera is used to resolve the backscattering echo of the transmitted laser beam.

Thus as in the case of conventional lidar technologies continuous wave light sources such as diode lasers can be employed for remote sensing instead of using complicated nanosecond pulse light sources. The SLidar system is also a robust and inexpensive system based on compact laser diodes and array detectors.

Law Enforcement

Lidar speed guns are used by the police to measure the speed of vehicles for speed limit enforcement purposes.

Military

Few military applications are known to be in place and are classified (like the lidar-based speed measurement of the AGM-129 ACM stealth nuclear cruise missile), but a considerable amount of research is underway in their use for imaging. Higher resolution systems collect enough detail to identify targets, such as tanks. Examples of military applications of lidar include the Airborne Laser Mine Detection System (ALMDS) for counter-mine warfare by Areté Associates.

A NATO report (RTO-TR-SET-098) evaluated the potential technologies to do stand-off detection for the discrimination of biological warfare agents. The potential technologies evaluated were Long-Wave Infrared (LWIR), Differential Scattering (DISC), and Ultraviolet Laser Induced Fluorescence (UV-LIF). The report concluded that : *Based upon the results of the lidar systems tested and discussed above, the Task Group recommends that the best option for the near-term (2008–2010) application of stand-off detection systems is UV LIF* . However, in the long-term, other techniques such as stand-off Raman spectroscopy may prove to be useful for identification of biological warfare agents.

Short-range compact spectrometric lidar based on Laser-Induced Fluorescence (LIF) would address the presence of bio-threats in aerosol form over critical indoor, semi-enclosed and outdoor venues like stadiums, subways, and airports. This near real-time capability would enable rapid detection of a bioaerosol release and allow for timely implementation of measures to protect occupants and minimize the extent of contamination.

The Long-Range Biological Standoff Detection System (LR-BSDS) was developed for the US Army to provide the earliest possible standoff warning of a biological attack. It is an airborne system carried by a helicopter to detect man-made aerosol clouds containing biological and chemical agents at long range. The LR-BSDS, with a detection range of 30 km or more, was fielded in June 1997. Five lidar units produced by the German company Sick AG were used for short range detection on Stanley, the autonomous car that won the 2005 DARPA Grand Challenge.

A robotic Boeing AH-6 performed a fully autonomous flight in June 2010, including avoiding obstacles using lidar.

Mining

Lidar is used in the mining industry for various tasks. The calculation of ore volumes is accomplished by periodic (monthly) scanning in areas of ore removal, then comparing surface data to the previous scan.

Lidar sensors may also be used for obstacle detection and avoidance for robotic mining vehicles such as in the Komatsu Autonomous Haulage System (AHS) used in Rio Tinto's Mine of the Future.

Physics and Astronomy

A worldwide network of observatories uses lidars to measure the distance to reflectors placed on the moon, allowing the position of the moon to be measured with mm precision and tests of general relativity to be done. MOLA, the Mars Orbiting Laser Altimeter, used a lidar instrument in a Mars-orbiting satellite (the NASA Mars Global Surveyor) to produce a spectacularly precise global topographic survey of the red planet.

In September, 2008, the NASA Phoenix Lander used lidar to detect snow in the atmosphere of Mars.

In atmospheric physics, lidar is used as a remote detection instrument to measure densities of certain constituents of the middle and upper atmosphere, such as potassium, sodium, or molecular nitrogen and oxygen. These measurements can be used to calculate temperatures. Lidar can also be used to measure wind speed and to provide information about vertical distribution of the aerosol particles.

At the JET nuclear fusion research facility, in the UK near Abingdon, Oxfordshire, lidar Thomson Scattering is used to determine Electron Density and Temperature profiles of the plasma.

Rock Mechanics

LiDAR has been widely used in rock mechanics for rock mass characterization and slope change detection. Some important geomechanical properties from the rock mass can be extracted from the 3D point clouds obtained by means of the LiDAR. Some of these properties are:

- Discontinuity orientation
- Discontinuity spacing and RQD
- Discontinuity aperture
- Discontinuity persistence
- Discontinuity roughness
- Water infiltration

Some of these properties have be used to assess the geomechanical quality of the rock mass through the RMR index. Moreover, as the orientations of discontinuities can be extracted using the existing methodologies, it is possible to assess the geomechanical quality of a rock slope through the SMR index. In addition to this, the comparison of different 3D point clouds from a slope acquired at different times allows to study the changes produced on the scene during this time interval as a result of rockfalls or any other landsliding processes.

Robotics

Lidar technology is being used in robotics for the perception of the environment as well as object classification. The ability of lidar technology to provide three-dimensional elevation maps of the terrain, high precision distance to the ground, and approach velocity can enable safe landing of robotic and manned vehicles with a high degree of precision. Refer to the Military section above for further examples.

Spaceflight

Lidar is increasingly being utilized for rangefinding and orbital element calculation of relative velocity in proximity operations and stationkeeping of spacecraft. Lidar has also been used for atmospheric studies from space. Short pulses of laser light beamed from a spacecraft can reflect off of tiny particles in the atmosphere and back to a telescope aligned with the spacecraft laser. By precisely timing the lidar 'echo,' and by measuring how much laser light is received by the telescope, scientists can accurately determine the location, distribution and nature of the particles. The result is a revolutionary new tool for studying constituents in the atmosphere, from cloud droplets to industrial pollutants, that are difficult to detect by other means."

Surveying

Airborne lidar sensors are used by companies in the remote sensing field. They can be used to create a DTM (Digital Terrain Model) or DEM (Digital Elevation Model); this is quite a common practice for larger areas as a plane can acquire 3–4 km wide swaths in a single flyover.

This TomTom mapping van is fitted with five lidars on its roof rack.

Greater vertical accuracy of below 50 mm can be achieved with a lower flyover, even in forests, where it is able to give the height of the canopy as well as the ground elevation. Typically, a GNSS receiver configured over a georeferenced control point is needed to link the data in with the WGS (World Geodetic System).

Transport

LiDAR has been used in the railroad industry to generate asset health reports for asset management and by departments of transportation to assess their road conditions. CivilMaps.com is a leading company in the field. Lidar has been used in adaptive cruise control (ACC) systems for automobiles. Systems such as those by Siemens and Hella use a lidar device mounted on the front of the vehicle, such as the bumper, to monitor the distance between the vehicle and any vehicle in front of it. In the event the vehicle in front slows down or is too close, the ACC applies the brakes to slow the vehicle. When the road ahead is clear, the ACC allows the vehicle to accelerate to a speed preset by the driver. Refer to the Military section above for further examples.

Wind Farm Optimization

Lidar can be used to increase the energy output from wind farms by accurately measuring wind speeds and wind turbulence. Experimental lidar systems can be mounted on the nacelle of a wind turbine or integrated into the rotating spinner to measure oncoming horizontal winds, winds in the wake of the wind turbine, and proactively adjust blades to protect components and increase power. Lidar is also used to characterise the incident wind resource for comparison with wind turbine power production to verify the performance of the wind turbine by measuring the wind turbine's power curve. Wind farm optimization can be considered a topic in *applied eolics*.

Solar Photovoltaic Deployment Optimization

Lidar can also be used to assist planners and developers in optimizing solar photovoltaic systems at the city level by determining appropriate roof tops and for determining shading losses. Recent works focus on buildings' facades solar potential estimation, or by incorporating more detailed shading losses by considering the influence from vegetation and larger surrounding terrain.

Video Games

Racing game iRacing features scanned tracks, resulting in bumps with millimeter precision in the in-game 3D mapping environment.

Other Uses

The video for the song "House of Cards" by Radiohead was believed to be the first use of real-time 3D laser scanning to record a music video. The range data in the video is not completely from a lidar, as structured light scanning is also used.

Alternative Technologies

Recent development of Structure From Motion (SFM) technologies allows delivering 3D images and maps based on data extracted from visual and IR photography. The elevation or 3D data is extracted using multiple parallel passes over mapped area, yielding both visual light image and 3D structure from the same sensor, which is often a specially chosen and calibrated digital camera.

SATELLITE LASER RANGING

In satellite laser ranging (SLR) a global network of observation stations measures the round trip time of flight of ultrashort pulses of light to satellites equipped with retroreflectors. This provides instantaneous range measurements of millimeter level precision which can be accumulated to provide accurate measurement of orbits and a host of important scientific data.

Satellite laser ranging is a proven geodetic technique with significant potential for important contributions to scientific studies of the earth/atmosphere/ocean system. It is the most accurate technique currently available to determine the geocentric position of an Earth satellite, allowing for the precise calibration of radar altimeters and separation of long-term instrumentation drift from secular changes in ocean topography.

Laser Ranging System of the geodetic observatory Wettzell, Bavaria.

Its ability to measure the variations over time in Earth's gravity field and to monitor motion of the station network with respect to the geocenter, together with the capability to monitor vertical motion in an absolute system, makes it unique for modeling and evaluating long-term climate change by:

- providing a reference system for post-glacial rebound, sea level and ice volume change
- determining the temporal mass redistribution of the solid earth, ocean, and atmosphere system
- monitoring the response of the atmosphere to seasonal variations in solar heating.

SLR provides a unique capability for verification of the predictions of the theory of general relativity, such as the frame-dragging effect.

SLR stations form an important part of the international network of space geodetic observatories, which include VLBI, GPS, DORIS and PRARE systems. On several critical missions, SLR has provided failsafe redundancy when other radiometric tracking systems have failed.

History

Laser ranging to a near-Earth satellite was first carried out by NASA in 1964 with the launch of the Beacon-B satellite. Since that time, ranging precision, spurred by scientific requirements, has improved by a factor of a thousand from a few meters to a few millimeters, and more satellites equipped with retroreflectors have been launched.

Several sets of retroreflectors were installed on Earth's moon as part of the American Apollo and Soviet Lunokhod space programs. These retroreflectors are also ranged on a regular basis (lunar laser ranging), providing a highly accurate measurement of the dynamics of the Earth/Moon system.

During the subsequent decades, the global satellite laser ranging network has evolved into a powerful source of data for studies of the solid Earth and its ocean and atmospheric systems. In addition, SLR provides precise orbit determination for spaceborne radar altimetre missions mapping the ocean surface (which are used to model global ocean circulation), for mapping volumetric changes in continental ice masses, and for land topography. It provides a means for subnanosecond global time transfer, and a basis for special tests of the Theory of General Relativity.

The International Laser Ranging Service was formed in 1998 by the global SLR community to enhance geophysical and geodetic research activities, replacing the previous CSTG Satellite and Laser Ranging Subcommission.

Uses of SLR Data

SLR data has provided the standard, highly accurate, long wavelength gravity field reference model which supports all precision orbit determination and provides the basis for studying temporal gravitational variations due to mass redistribution. The height of the geoid has been determined to less than ten centimeters at long wavelengths less than 1500 km.

SLR provides mm/year accurate determinations of tectonic drift station motion on a global scale in a geocentric reference frame. Combined with gravity models and decadal changes in Earth rotation, these results contribute to modeling of convection in the Earth's mantle by providing constraints on related Earth interior processes. The velocity of the fiducial station in Hawaii is 70 mm/year and closely matches the rate of the background geophysical model.

REFERENCES

1. "LIDAR—Light Detection and Ranging—is a remote sensing method used to examine the surface of the Earth". NOAA. Archived from the original on June 4, 2013. Retrieved June 4, 2013.
2. Oxford English Dictionary. 2013. p. Entry for "lidar".
3. James Ring, "The Laser in Astronomy." p. 672–673, New Scientist Jun 20, 1963
4. Cracknell, Arthur P.; Hayes, Ladson (2007) . Introduction to Remote Sensing (2 ed.). London: Taylor and Francis. ISBN 0-8493-9255-1. OCLC 70765252.
5. Goyer, G. G.; R. Watson (September 1963). "The Laser and its Application to Meteorology". Bulletin of the American Meteorological Society. 44 (9): 564–575 .

6. "New Artillery Against Smog: TV and Lidar" Popular Mechanics, April 1970, p. 104.
7. "Light Detection and Ranging". Ngs.noaa.gov. 2011-04-15. Retrieved 2016-02-22.
8. US 20090273770 patent
9. "CLICK". Lidar.cr.usgs.gov.. 2015-09-16. Retrieved 2016-02-22.
10. Carter, Jamie; Keil Schmid; Kirk Waters; Lindy Betzhold; Brian Hadley; Rebecca Mataosky; Jennifer Halleran (2012). "Lidar 101: An Introduction to Lidar Technology, Data, and Applications." (NOAA) Coastal Services Center" (PDF). Coast.noaaa.gov. p. 14. Retrieved 2016-02-22.
11. 'Experimental Advanced Advanced Research Lidar', USGS.gov. Retrieved 8 August 2007.
12. Amzajerdian, Farzin; Pierrottet, Diego F.; Petway, Larry B.; Hines, Glenn D.; Roback, Vincent E. "Lidar Systems for Precision Navigation and Safe Landing on Planetary Bodies". Langel Research Center. NASA. Retrieved May 24, 2011.
13. Medina, Antonio. "Three Dimensional Camera and Rangefinder". January 1992. United States Patent 5081530.
14. Medina A, Gayá F, Pozo F (2006). "Compact laser radar and three-dimensional camera". J. Opt. Soc. Am. A. 23: 800–805. Bibcode:2006JOSAA..23..800M. doi:10.1364/josaa.23.000800.
15. Strauss, C. E. M. (1994). "Synthetic-array heterodyne detection: a single-element detector acts as an array". Opt. Lett. 19: 1609–1611. Bibcode:1994OptL...19.1609S. doi:10.1364/ol.19.001609.
16. Talbot, David (2014-02-13). "New Optical Chip Will Sharpen Military and Archeological Aerial Imaging | MIT Technology Review". Technologyreview.com. Retrieved 2014-02-17.
17. Vosselman, George; Maas, Hans-Gerd (2012). Airborne and terrestrial laser scanning. Whittles Publishing. ISBN 978-1-904445-87-6.
18. Doneus, M.; Miholjek, I.; Mandlburger, G.; Doneus, N.; Verhoeven, G.; Briese, Ch.; Pregesbauer, M. "Airborne laser bathymetry for documentation of submerged archaeological sites in shallow water". ISPRS - International Archives of the Photogrammetry, Remote Sensing and Spatial Information Sciences. XL-5/W5: 99–107. Bibcode:2015ISPArXL55...99D. doi:10.5194/isprsarchives-xl-5-w5-99-2015.
19. Chiu, Cheng-Lung; Fei, Li-Yuan; Liu, Jin-King; Wu, Ming-Chee. "National Airborne Lidar Mapping and Examples for applications in deep seated landslides in Taiwan". Geoscience and Remote Sensing Symposium (IGARSS), 2015 IEEE International. ISSN 2153-7003.
20. Yuan, Zeng; Yujin, Zhao; Dan, Zhao; Bingfang, Wu. "Forest Biodiversity mapping using airborne and hyper-spectral data". Geoscience and Remote Sensing Symposium (IGARSS), 2016 IEEE International,. ISSN 2153-7003.
21. C. Wilson, Jerry. "Using Airborne hydrographic lidar to support mapping of

California's waters". OCEANS 2008 - MTS/IEEE Kobe Techno-Ocean, ISBN 978-1-4244-2126-8.

22. Tang, Lina; Shao, Guofan (2015-06-21). "Drone remote sensing for forestry research and practices". Journal of Forestry Research. 26 (4): 791–797. doi:10.1007/s11676-015-0088-y. ISSN 1007-662X.
23. Lee, Sang-Mook; Joon Im, Jeong; Lee, Bo-Hee; Leonessa, Alexander; Kurdila, Andrew. "A real time grid-map generation and object classification for ground based 3D lidar data using image analysis techniques". Image Processing (ICIP), 2010 17th IEEE International Conference on image processing,. ISSN 1522-4880.
24. Takagi, Kiyokazu; Morikawa, Katsuhiro; Ogawa, Takashi; Saburi, Makoto (2006). "Road Environment Recognition Using On-vehicle LIDAR". IEEE Intelligent Vehicles Symposium. ISBN 4-901122-86-X.
25. "ARS Study Helps Farmers Make Best Use of Fertilizers". USDA Agricultural Research Service. June 9, 2010.
26. Weiss, Ulrich; Biber, Peter; Laible, Stefan; Bohlmann, Karsten; Zell, Andreas (2010). "Plant Species Classification using a 3D LIDAR Sensor and Machine Learning". Ninth International Conference on Machine Learning and Applications (ICMLA). ISBN 978-1-4244-9211-4.
27. "EID; crater beneath canopy". Unb.ca. 2013-02-18. Retrieved 2013-05-06.
28. John Nobel Wilford (2010-05-10). "Mapping Ancient Civilization, in a Matter of Days". New York Times. Retrieved 2010-05-11.
29. The Light Fantastic: Using airborne lidar in archaeological survey. English Heritage. 2010. p. 45.
30. Stephanie Pappas (May 15, 2013). "Ruins of Lost City May Lurk Deep in Honduras Rain Forest". Live Science. Retrieved May 15, 2013.
31. "Jungle surrenders its lost city". Smh.com.au. Retrieved 2016-02-22.
32. "Rediscovering the lost archaeological landscape of southern New England using airborne light detection and ranging (LiDAR)". Journal of Archaeological Science. 43: 9–20. doi:10.1016/j.jas.2013.12.004. Retrieved 2016-02-22.
33. Edwin Cartlidge. "Lasers Unearth Lost 'Agropolis' of New England | Science | AAAS". News.sciencemag.org. Retrieved 2016-02-22.
34. ""Lost" New England Revealed by High-Tech Archaeology". News.nationalgeographic.com. 2014-01-03. Retrieved 2016-02-22.
35. Douglas Preston (2 Mar 2015). "Lost City Discovered in the Honduran Rain Forest". National Geographic. Retrieved 3 March 2015.
36. D.H. Evans,R.J. Fletcher et al., Uncovering archaeological landscapes at Angkor using lidar, PNAS, vol. 110, no. 31, 12595–12600, doi: 10.1073/pnas.1306539110| http://www.pnas.org/content/110/31/12595.full
37. By Steve Taranovich, EDN. "Autonomous automotive sensors: How processor algorithms get their inputs." July 5, 2016. Retrieved August 9, 2016.

38. "Filipino turns ordinary car into autonomous vehicle - Motioncars | Motioncars". Motioncars.inquirer.net. 2015-05-25. Retrieved 2016-02-22.

39. "Sick Senor Intelligence product portfolio". 2014-11-12. Retrieved 2014-11-12.

40. "Hokuyo scanning range finder". 2014-11-12. Retrieved 2014-11-12.

41. "Autonomous Solutions Forecast 3D Laser System". 2013-01-24. Retrieved 2013-05-06.

42. "Products". Velodynelidar.com. Retrieved 2016-02-22.

43. Lindner, Philipp; Wanielik, Gerd (2009). "3D LIDAR Processing for Vehicle Safety and Environment Recognition". IEEE Workshop on Computational Intelligence in Vehicles and Vehicular Systems. ISBN 978-1-4244-2770-3.

44. Hwang, Soonmin; Kim, Namil; Choi, Yukyung; Lee, Seokju; So Kweon, In (2016). "Fast Multiple Objects Detection and Tracking Fusing Color Camera and 3D LIDAR for Intelligent Vehicles". 13th International Conference on Ubiquitous Robots and Ambient Intelligence (URAI). ISBN 978-1-5090-0821-6.

45. Cheng, Ming-Ming; Zhang, Ziming; Lin, Wen-Yan; Torr, Philip (2014). "BING: Binarized normed gradients for objectness estimation at 300fps". IEEE Conference on Computer Vision and Pattern Recognition (CVPR). ISBN 978-1-4799-5118-5.

46. Girshick, Ross (2015). "Fast R-CNN". IEEE International Conference on Computer Vision (ICCV). ISBN 978-1-4673-8391-2.

47. E. Johnson, Andrew; Hebert, Martial (1999). "Using spin images for efficient object recognition in cluttered 3D scenes". IEEE Transactions on Pattern Analysis and Machine Intelligence (TPAMI). ISSN 0162-8828.

48. Xu, Philippe; Davoine, Franck; Denoeux, Thierry (September 2014). "Evidential combination of pedestrian detectors". Proceedings British Machine Vision Conference, BMVA Press.

49. Zhou, Kun; Wang, Xiqin; Tomizukat, Masayoshi; Zhang, Wei-Bin; Chant, Ching-Yao (2002). "A New Manoeuvring Target Tracking Algorithm with Input Estimation". Proceedings of the 2002 American Control Conference. ISBN 0-7803-7298-0.

50. Y. Hata, Alberto; F. Wolf, Denis. "Feature Detection for Vehicle Localization in Urban Environments Using a Multilayer LIDAR". IEEE Transaction on Intelligent Transportation System. 17 (2). ISSN 1558-0016.

51. https://www.theguardian.com/technology/2015/sep/07/hackers-trick-self-driving-cars-lidar-sensor

52. "Lidar Links of Use in Mapping". Lidarbasemaps.org. Retrieved 2016-02-22.

53. Councillor Quarterly, Summer 2007 Volume 6 Issue 3

54. Hughes, M. W.; Quigley, M. C; van Ballegooy, S.; Deam, B. L.; Bradley, B. A.; Hart, D. E. (2015). "The sinking city: Earthquakes increase flood hazard in Christchurch, New Zealand". GSA Today. 25 (3): 4–10. Retrieved 2016-02-22.

55. Rabatel, Antoine; Deline, Philip; Jaillet, Ste'phane; Ravanel, Ludovic (28 May 2008). "Rock falls in high-alpine rock walls quantified by terrestrial lidar

measurements: A case study in the Mont Blanc area". Geophysical Research Letters. 35. Bibcode:2008GeoRL..3510502R. doi:10.1029/2008GL033424. Retrieved 4 December 2015.

56. "LIDAR shows where earthquake risks are highest". Seattlepi.com. 2001-04-17. Retrieved 2016-02-22.
57. 'Mount Saint Helens LIDAR Data', Washington State Geospatial Data Archive (September 13, 2006). Retrieved 8 August 2007.
58. 'Airborne Topographic Mapper', NASA.gov. Retrieved 8 August 2007.
59. "Lecture 14 : Principles of active remote sensing: Lidars and lidar sensing of aerosols, gases and clouds." (PDF). Laser-distance-measurer.com. Retrieved 2016-02-22.
60. Li,, T. (2011). "Middle atmosphere temperature trend and solar cycle revealed by long-term Rayleigh lidar observations". J. Geophys. Res. 116. Bibcode:2011JGRD..116.0P05L. doi:10.1029/2010jd015275.
61. Thomas D. Wilkerson, Geary K. Schwemmer, and Bruce M. Gentry. LIDAR Profiling of Aerosols, Clouds, and Winds by Doppler and Non-Doppler Methods, NASA International H2O Project (2002).
62. 'Earth Explorers: ADM-Aeolus', ESA.org (European Space Agency, 6 June 2007). Retrieved 8 August 2007.
63. 'Doppler lidar gives Olympic sailors the edge', Optics.org (3 July, 2008). Retrieved 8 July 2008.
64. Clive, P. J. M., The emergence of eolics, TEDx University of Strathclyde (2014). Retrieved 9 May 2014.
65. Koch, Grady J.; Barnes, Bruce W; Petros, Mulugeta; Beyon, Jeffrey Y; Amzajerdian, Farzin; Yu, Jirong; Davis, Richard E; Ismail, Syed; Vay, Stephanie; Kavaya, Michael J; Singh, Upendra N. "Coherent Differential Absorption Lidar Measurements of CO2". Applied Optics. 43 (26): 5092–5099. Bibcode:2004ApOpt..43.5092K. doi:10.1364/AO.43.005092.
66. Abshire, James B.; Ramanathan, Anand; Riris, Haris; Mao, Jianping; Allan, Graham R.; Hasselbrack, William E.; Weaver, Clark J.; Browell, Edward V. (2013-12-30). "Airborne Measurements of CO2 Column Concentration and Range Using a Pulsed Direct-Detection IPDA Lidar". Remote Sensing. 6 (1): 443–469. doi:10.3390/rs6010443.
67. Campbell, Joel F. "Nonlinear swept frequency technique for CO2 measurements using a CW laser system". Applied Optics. 52 (13): 3100–3107. arXiv:1303.4933Freely accessible. Bibcode:2013ApOpt..52.3100C. doi:10.1364/AO.52.003100.
68. Dobler, Jeremy T.; Harrison, F. Wallace; Browell, Edward V.; Lin, Bing; McGregor, Doug; Kooi, Susan; Choi, Yonghoon; Ismail, Syed. "Atmospheric CO2 column measurements with an airborne intensity-modulated continuous wave 1.57 μm fiber laser lidar. Applied Optics. 52 (12): 2874–2892. Bibcode:2013ApOpt..52.2874D. doi:10.1364/AO.52.002874.

69. Riris, Haris; Numata, Kenji; Li, Steve; Wu, Stewart; Ramanathan, Anand; Dawsey, Martha; Mao, Jianping; Kawa, Randolph; Abshire, James B. (2012-12-01). Airborne measurements of atmospheric methane column abundance using a pulsed integrated-path differential absorption lidar. Applied Optics. 51 (34): 8296. Bibcode:2012ApOpt..51.8296R. doi:10.1364/AO.51.008296. ISSN 1539-4522.
70. Grant, W. B., Lidar for atmospheric and hydrospheric studies, in Tunable Laser Applications, 1st Edition, Duarte, F. J. Ed. (Marcel Dekker, New York, 1995) Chapter 7.
71. Mei, Liang. Remote sensing of atmospheric aerosol and gas using scheimpflug lidar based on diode lasers. Progress in Electromagnetic Research Symposium (PIERS),. ISBN 978-1-5090-6094-8.
72. Gonglach, Matt. How Police Laser Guns Work.
73. Archived September 4, 2011, at the Wayback Machine.
74. NATO Laser Based Stand-Off Detection of biological Agents. Rta.nato.int. Retrieved 2013-05-06.
75. Short-Range Bioaerosol Threat Detection Sensor (SR-BioSpectra). Ino.ca. Retrieved 2013-05-06.
76. Defense & Security Intelligence & Analysis: IHS Jane›s | IHS. Articles.janes.com. Retrieved 2016-02-22.
77. Spice, Byron. Researchers Help Develop Full-Size Autonomous Helicopter Carnegie Mellon, 6 July 2010. Retrieved: 19 July 2010.
78. Koski, Olivia. In a First, Full-Sized Robo-Copter Flies With No Human Help Wired, 14 July 2010. Retrieved: 19 July 2010.
79. Volume Measuring. 3dlasermapping.com. Retrieved 2014-02-17.
80. Modular Mining Systems#Autonomous Haulage Systems
81. NASA. ‹NASA Mars Lander Sees Falling Snow, Soil Data Suggest Liquid Past› NASA.gov (29 September 2008). Retrieved 9 November 2008.
82. Cloud-Aerosol Lidar with Orthogonal Polarization (CALIOP). NASA. Retrieved 16 August 2015.
83. CW Gowers. ‹ Focus On : Lidar-Thomson Scattering Diagnostic on JET› JET. EFDA.org (undated). Retrieved 8 August 2007. Archived September 18, 2007, at the Wayback Machine.
84. Riquelme, A.J., Abellán, A., Tomás, R. & Jaboyedoff, M. 2014. A new approach for semi-automatic rock mass joints recognition from 3D point clouds. Computers & Geosciences, 68, 38–52, doi: 10.1016/j.cageo.2014.03.014.
85. Gigli, G. & Casagli, N. 2011. Semi-automatic extraction of rock mass structural data from high resolution LIDAR point clouds. International Journal of Rock Mechanics and Mining Sciences, 48, 187–198, doi:10.1016/j.ijrmms.2010.11.009.
86. Slob, S. 2010. Automated rock mass characterization using 3D terrestrial laser scanner, Technical University of Delf.
87. Riquelme, A.J., Abellán, A. & Tomás, R. 2015. Discontinuity spacing analysis

in rock masses using 3D point clouds. Engineering Geology, 195, 185–195, doi: http://dx.doi.org/10.1016/j.enggeo.2015.06.009.

88. Sturzenegger, M. & Stead, D. 2009. Close-range terrestrial digital photogrammetry and terrestrial laser scanning for discontinuity characterization on rock cuts. Engineering Geology, 106, 163–182, doi:10.1016/j.enggeo.2009.03.004.

89. Riquelme, Adrián J.; Tomás, Roberto; Abellán, Antonio (2016-04-01). Characterization of rock slopes through slope mass rating using 3D point clouds. International Journal of Rock Mechanics and Mining Sciences. 84: 165–176. doi:10.1016/j.ijrmms.2015.12.008.

90. Abellán, A., Oppikofer, T., Jaboyedoff, M., Rosser, N.J., Lim, M. & Lato, M.J. 2014. Terrestrial laser scanning of rock slope instabilities. Earth Surface Processes and Landforms, 39, 80–97, doi: 10.1002/esp.3493.

91. Abellán, A., Vilaplana, J.M. & Martínez, J. 2006. Application of a long-range Terrestrial Laser Scanner to a detailed rockfall study at Vall de Núria (Eastern Pyrenees, Spain). Engineering Geology, 88, 136–148, doi:10.1016/j.enggeo.2006.09.012.

92. IfTAS. Iftas.de. Retrieved 2013-05-06.

93. Amzajerdian, Farzin; Pierrottet, Diego F.; Petway, Larry B.; Hines, Glenn D.; Roback, Vincent E. Lidar Systems for Precision Navigation and Safe Landing on Planetary Bodies. Langley Research Center. NTRS. Retrieved May 24, 2011.

94. NASA - Lidar In-space Technology Experiment (LITE). Nasa.gov. 2011-08-25. Retrieved 2013-05-06.

95. D.M. Winker ; R.H. Couch ; M.P. McCormick (2011-09-27). IEEE Xplore - An overview of LITE: NASA›s Lidar In-space Technology Experiment (PDF). Ieeexplore.ieee.org. doi:10.1109/5.482227. Retrieved 2013-05-06.

96. Szulwic, Jakub; Burdziakowski, Paweł; Janowski, Artur; Przyborski, Marek; Tysiąc, Paweł; Wojtowicz, Aleksander; Kholodkov, Arthem; Matysik, Krzysztof; Matysik, Maciej. Maritime Laser Scanning as the Source for Spatial Data. Polish Maritime Research. 22 (4). doi:10.1515/pomr-2015-0064.

97. CivilMaps.com accelerating road and pavement condition surveys. Civil Maps. 2015-03-15. Retrieved 2015-03-15.

98. Bumper-mounted lasers. Sciencedaily.com. 2007-02-27. Retrieved 2013-05-06.

99. Clive, P. J. M., Windpower 2.0: technology rises to the challenge Environmental Research Web, 2008. Retrieved: 9 May 2014.

100. Mikkelsen, Torben; et al. (October 2007). 12MW Horns Rev Experiment (PDF). Risoe. Retrieved 2010-04-25.

101. Smarting from the wind. The Economist. 2010-03-04. Retrieved 2010-04-25.

102. The world›s first control of a wind turbine with a nacelle-based Lidar system. Corporate Communications University of Stuttgart. 2012-06-05. Retrieved 2014-04-12.

103. Andrew K. Scholbrock et al. Field Testing LIDAR Based Feed-Forward Controls

on the NREL Controls Advanced Research Turbine National Renewable Energy Laboratory Data Base, 12 April 2014. Retrieved: 12 April 2014.

104. Mikkelsen, Torben & Hansen, Kasper Hjorth et al. Lidar wind speed measurements from a rotating spinner Danish Research Database & Danish Technical University, 20 April 2010. Retrieved: 25 April 2010.

105. Asimakopolous, M., Clive, P. J. M., More, G., and Boddington, R., Offshore compression zone measurement and visualisation European Wind Energy Association Annual Conference, 2014. Retrieved: 9 May 2014.

106. Gallacher, D., and More, G., Lidar measurements and visualisation of turbulence and wake decay length European Wind Energy Association Annual Conference, 2014. Retrieved: 9 May 2014.

107. Clive, P. J. M., et al., Offshore power curve tests for onshore costs: a real world case study European Wind Energy Association Annual Conference, 2014. Retrieved: 9 May 2014.

108. Clive, P. J. M., Offshore power performance assessment for onshore costs DEWEK (Deutsche Windenergie Konferenz), 2012. Retrieved: 9 May 2014.

109. Jochem, Andreas; Höfle Bernhard; Rutzinger Martin; Pfeifer Norbert (2009). Automatic roof plane detection and analysis in airborne lidar point clouds for solar potential assessment. Sensors. 9 (7): 5241–5262. doi:10.3390/s90705241.

110. Nguyen, Ha T.; Pearce, Joshua M.; Harrap, Rob; Barber, Gerald (2012). The Application of LiDAR to Assessment of Rooftop Solar Photovoltaic Deployment Potential on a Municipal District Unit. Sensors. 12: 4534–4558. doi:10.3390/s120404534.

111. Nguyen, Ha T.; Pearce, Joshua M. (2012). Incorporating shading losses in solar photovoltaic potential assessment at the municipal scale. Solar Energy. 86 (5): 1245–1260. Bibcode:2012SoEn...86.1245N. doi:10.1016/j.solener.2012.01.017.

112. Jochem, Andreas; Höfle Bernhard; Rutzinger Martin (2011). Extraction of vertical walls from mobile laser scanning data for solar potential assessment. Remote Sensing. 3 (4): 650–667. Bibcode:2011RemS....3..650J. doi:10.3390/rs3030650.

113. Lukač, Niko; Žlaus Danijel; Seme Seme; Žalik Borut; Štumberger Gorazd (2013). Rating of roofs› surfaces regarding their solar potential and suitability for PV systems, based on LiDAR data. Applied Energy. 102: 803–812. doi:10.1016/j.apenergy.2012.08.042.

114. Nick Parish (2008-07-13). From OK Computer to Roll computer: Radiohead and director James Frost make a video without cameras. Creativity.

CHAPTER 7

Aerial Photography

Chapter Contents

- **Introduction**
- **History**
- **Uses**
- **Satellite Imagery**
- **History**

An aerial photograph, in broad terms, is any photograph taken from the air. Normally, air photos are taken vertically from an aircraft using a highly-accurate camera. There are several things you can look for to determine what makes one photograph different from another of the same area including type of film, scale, and overlap. Other important concepts used in aerial photography are stereoscopic coverage, fiducial marks, focal length, roll and frame numbers, and flight lines and index maps. The following material will help you understand the fundamentals of aerial photography by explaining these basic technical concepts.

INTRODUCTION

Aerial photography is the taking of photographs of the ground from an elevated/direct-down position. Usually the camera is not supported by a ground-based structure. Platforms for aerial photography include fixed-wing aircraft, helicopters, unmanned aerial vehicles (UAVs or "drones"), balloons, blimps and dirigibles, rockets, pigeons, kites, parachutes, stand-alone telescoping and vehicle-mounted poles. Mounted cameras may be triggered remotely or automatically; hand-held photographs may be taken by a photographer.

Aerial photography should not be confused with air-to-air photography, where one or more aircraft are used as chase planes that "chase" and photograph other aircraft in flight.

An aerial photograph using a drone of Westerhevers and Lighthouse, Germany.

Air photo of a military target used to evaluate the effect of bombing.

HISTORY

Early History

Honoré Daumier, "Nadar élevant la Photographie à la hauteur de l›Art» (Nadar elevating Photography to Art), published in *Le Boulevard*, May 25, 1862.

Aerial photography was first practiced by the French photographer and balloonist Gaspard-Félix Tournachon, known as "Nadar", in 1858 over Paris, France. However, the photographs he produced no longer exist and therefore the earliest surviving aerial photograph is titled 'Boston, as the Eagle and the Wild Goose See It.' Taken by James Wallace Black and Samuel Archer King on October 13, 1860, it depicts Boston from a height of 630m.

Antique postcard using kite photo technique.

Kite aerial photography was pioneered by British meteorologist E.D. Archibald in 1882. He used an explosive charge on a timer to take photographs from the air. Frenchman Arthur Batut began using kites for photography in 1888, and wrote a book on his methods in 1890. Samuel Franklin Cody developed his advanced 'Man-lifter War Kite' and succeeded in interesting the British War Office with its capabilities.

The first use of a motion picture camera mounted to a heavier-than-air aircraft took place on April 24, 1909 over Rome in the 3:28 silent film short, *Wilbur Wright und seine Flugmaschine*.

World War I

The use of aerial photography rapidly matured during the war, as reconnaissance aircraft were equipped with cameras to record enemy movements and defences. At the start of the conflict, the usefulness of aerial photography was not fully appreciated, with reconnaissance being accomplished with map sketching from the air.

Germany adopted the first aerial camera, a Görz, in 1913. The French began the war with several squadrons of Blériot observation aircraft equipped with cameras for reconnaissance. The French Army developed procedures for getting prints into the hands of field commanders in record time.

Giza pyramid complex, photographed from Eduard Spelterini's balloon on November 21, 1904

Frederick Charles Victor Laws started aerial photography experiments in 1912 with No.1 Squadron of the Royal Flying Corps (later No. 1 Squadron RAF), taking photographs from the British dirigible *Beta*. He discovered that vertical photos taken with 60% overlap could be used to create a stereoscopic effect when viewed in a stereoscope, thus creating a perception of depth that could aid in cartography and in intelligence derived from aerial images. The Royal Flying Corps recon pilots began to use cameras for recording their observations in 1914 and by the Battle of Neuve Chapelle in 1915, the entire system of German trenches was being photographed. In 1916 the Austro-Hungarian Monarchy made vertical camera axis aerial photos above Italy for map-making.

A German observation plane, the Rumpler Taube.

The first purpose-built and practical aerial camera was invented by Captain John Moore-Brabazon in 1915 with the help of the Thornton-Pickard company, greatly enhancing the efficiency of aerial photography. The camera was inserted into the floor of the aircraft and could be triggered by the pilot at intervals. Moore-Brabazon also pioneered the incorporation of stereoscopic techniques into aerial photography, allowing the height of objects on the landscape to be discerned by comparing photographs taken at different angles.

By the end of the war aerial cameras had dramatically increased in size and focal power and were used increasingly frequently as they proved their pivotal military worth; by 1918 both sides were photographing the entire front twice a day, and had taken over half a million photos since the beginning of the conflict. In January 1918, General Allenby used five Australian pilots from No. 1 Squadron AFC to photograph a 624 square miles (1,620 km^2) area in Palestine as an aid to correcting and improving maps of the Turkish front. This was a pioneering use of aerial photography as an aid for cartography. Lieutenants Leonard Taplin, Allan Runciman Brown, H. L. Fraser, Edward Patrick Kenny, and L. W. Rogers photographed a block of land stretching from the Turkish front lines 32 miles (51 km) deep into their rear areas. Beginning 5 January, they flew with a fighter escort to ward off enemy fighters. Using Royal Aircraft Factory BE.12 and Martinsyde airplanes, they not only overcame enemy air attacks, but also had to contend with 65 mph (105 km/h) winds, antiaircraft fire, and malfunctioning equipment to complete their task.

Commercial Aerial Photography

The first commercial aerial photography company in the UK was Aerofilms Ltd, founded by World War I veterans Francis Wills and Claude Graham White in 1919. The company soon expanded into a business with major contracts in Africa and Asia as well as in the UK. Operations began from the Stag Lane Aerodrome at Edgware, using the aircraft of the London Flying School. Subsequently the Aircraft Manufacturing Company (later the De Havilland Aircraft Company), hired an Airco DH.9 along with pilot entrepreneur Alan Cobham.

New York City 1930, aerial photograph of Fairchild Aerial Surveys Inc.

From 1921, Aerofilms carried out vertical photography for survey and mapping purposes. During the 1930s, the company pioneered the science of photogrammetry (mapping from aerial photographs), with the Ordnance Survey amongst the company's clients.

Another successful pioneer of the commercial use of aerial photography was the American Sherman Fairchild who started his own aircraft firm Fairchild Aircraft to develop and build specialized aircraft for high altitude aerial survey missions. One Fairchild aerial survey aircraft in 1935 carried unit that combined two synchronized cameras, and each camera having five six inch lenses with a ten-inch lenses and took photos from 23,000 feet. Each photo covered 225 square miles. One of its first government contracts was an aerial survey of New Mexico to study soil erosion. A year later, Fairchild introduced a better high altitude camera with nine-lens in one unit that could take a photo of 600 square miles with each exposure from 30,000 feet.

World War II

In 1939, Sidney Cotton and Flying Officer Maurice Longbottom of the RAF were among the first to suggest that airborne reconnaissance may be a task better suited to fast, small aircraft which would use their speed and high service ceiling to avoid detection and interception. Although this seems obvious now, with modern reconnaissance tasks performed by fast, high flying aircraft, at the time it was radical thinking.

Sidney Cotton's Lockheed 12A, in which he made a high-speed reconnaissance flight in 1940.

They proposed the use of Spitfires with their armament and radios removed and replaced with extra fuel and cameras. This led to the development of the Spitfire PR variants. Spitfires proved to be extremely successful in their reconnaissance role and there were many variants built specifically for that purpose. They served initially with what later became No. 1 Photographic Reconnaissance Unit (PRU). In 1928, the RAF developed an electric heating system for the aerial camera. This allowed reconnaissance aircraft to take pictures from very high altitudes without the camera parts freezing. Based at RAF Medmenham, the collection and interpretation of such photographs became a considerable enterprise.

Cotton's aerial photographs were far ahead of their time. Together with other members of the 1 PRU, he pioneered the techniques of high-altitude, high-speed stereoscopic photography that were instrumental in revealing the locations of many crucial military and intelligence targets. According to R.V. Jones, photographs were used to establish the size and the characteristic launching mechanisms for both the V-1 flying bomb and the V-2 rocket. Cotton also worked on ideas such as a prototype specialist reconnaissance aircraft and further refinements of photographic equipment. At the peak, the British flew over 100 reconnaissance flights a day, yielding 50,000 images per day to interpret. Similar efforts were taken by other countries.

USES

Aerial photography is used in cartography (particularly in photogrammetric surveys, which are often the basis for topographic maps), land-use planning, archaeology, movie production, environmental studies, power line inspection, surveillance, commercial advertising, conveyancing, and artistic projects. An example of how aerial photography is used in the field of archaeology is the mapping project done at the site Angkor Borei in Cambodia from 1995-1996.

Abalone point .. Irvine Cove, Laguna Beach an example of low-altitude aerial photography

Using aerial photography, archaeologists were able to identify archaeological features, including 112 water features (reservoirs, artificially constructed pools and natural ponds) within the walled site of Angkor Borei. In the United States, aerial photographs are used in many Phase I Environmental Site Assessments for property analysis.

Platforms

Aircraft

Full-size, manned aircraft are prohibited from flights under 1000 feet, over congested areas and 500 feet above more sparsely populated locations.

Radio-Controlled Model Aircraft

Advances in radio controlled models have made it possible for model aircraft to conduct low-altitude aerial photography. This had benefited real-estate advertising, where commercial and residential properties are the photographic subject when in 2014 the US Federal

Communications Commission, issued an order banning the use of "Drones" in any commercial application related to photographs for use in real estate advertisement's. This ban has since been lifted, as the FAA Part 107 regulations for small UAS became effective on August 29, 2016.

A drone carrying a camera for aerial photography

Two drones that can be used to take aerial photographs

Small scale model aircraft offer increased photographic access to these previously restricted areas. Miniature vehicles do not replace full size aircraft, as full size aircraft are capable of longer flight times, higher altitudes, and greater equipment payloads. They are, however, useful in any situation in which a full-scale aircraft would be dangerous to operate. Examples would include the inspection of transformers atop power transmission lines and slow, low-level flight over agricultural fields, both of which can be accomplished by a large-scale radio controlled helicopter. Professional-grade, gyroscopically stabilized camera platforms are available for use under such a model; a large model helicopter with a 26cc gasoline engine can hoist a payload of approximately seven kilograms (15 lbs). In addition to gyroscopically stabilized footage, the use of RC copters as reliable aerial photography tools increased with the integration of FPV (first-person-view) technology. Many radio-controlled aircraft are now capable of utilizing Wi-Fi to stream live video from the aircraft's camera back to the pilot's ground station.

Regulations

Australia

In Australia Civil Aviation Safety Regulation 101 (CASR 101) allows for commercial use of radio control aircraft. Under these regulations radio controlled unmanned aircraft for commercial are referred to as Unmanned Aircraft Systems (UAS), where as radio controlled aircraft for recreational purposes are referred to as model aircraft. Under CASR 101, businesses/persons operating radio controlled aircraft commercially are required to hold an operator certificate, just like manned aircraft operators. Pilots of radio controlled aircraft operating commercially are also required to be licensed by the Civil Aviation Safety Authority (CASA). Whilst a small UAS and model aircraft may actually be identical, unlike model aircraft, a UAS may enter controlled airspace with approval, and operate within close proximity to an aerodrome.

Due to a number of illegal operators in Australia making false claims of being approved, CASA maintains and publishes a list of approved UAS operators.

However, CASA has modified the regulations and from the 29th of September 2016 drones under 2 kg may be operated for commercial purposes. This makes a lot of sense in creating a more competitive market place for basic services such as aerial photography in the real estate industry and farming.

United States

2006 FAA regulations grounding all commercial RC model flights have been upgraded to require formal FAA certification before permission is granted to fly at any altitude in the US.

June 25, 2014, The FAA, in ruling 14 CFR Part 91 [Docket No. FAA–2014–0396] "Interpretation of the Special Rule for Model Aircraft", banned the commercial use of unmanned aircraft over U.S. airspace. On September 26, 2014, the FAA began granting the right to use drones in aerial filmmaking. Operators are required to be licensed pilots and must keep the drone in view at all times. Drones cannot be used to film in areas where people might be put at risk.

The FAA Modernization and Reform Act of 2012 established, in Section 336, a special rule for model aircraft. In Section 336, Congress confirmed the FAA's long-standing position that model aircraft are aircraft. Under the terms of the Act, a model aircraft is defined as "an unmanned aircraft" that is "(1) capable of sustained flight in the atmosphere; (2) flown within visual line of sight of the person operating the aircraft; and (3) flown for hobby or recreational purposes."

Because anything capable of being viewed from a public space is considered outside the realm of privacy in the United States, aerial photography may legally document features and occurrences on private property.

The FAA can pursue enforcement action against persons operating model aircraft who endanger the safety of the national airspace system. Public Law 112–95, section 336(b).

TYPES

Oblique

Photographs taken at an angle are called *oblique photographs*. If they are taken from a low angle earth surface–aircraft, they are called *low oblique* and photographs taken from a high angle are called *high* or *steep oblique*.

Oblique Aerial Photo

An aerial photographer prepares continuous oblique shooting in a Cessna 206

Vertical

Vertical Orientation Aerial Photo

Vertical photographs are taken straight down. They are mainly used in photogrammetry and image interpretation. Pictures that will be used in photogrammetry are traditionally taken with special large format cameras with calibrated and documented geometric properties.

Combinations

Aerial photographs are often combined. Depending on their purpose it can be done in several ways, of which a few are listed below.

- Panoramas can be made by stitching several photographs taken with one hand held camera.
- In pictometry five rigidly mounted cameras provide one vertical and four low oblique pictures that can be used together.
- In some digital cameras for aerial photogrammetry images from several imaging elements, sometimes with separate lenses, are geometrically corrected and combined to one image in the camera.

Orthophotos

Vertical photographs are often used to create orthophotos, alternatively known as orthophotomaps, photographs which have been geometrically "corrected" so as to be usable as a map. In other words, an orthophoto is a simulation of a photograph taken from an infinite distance, looking straight down to nadir. Perspective must obviously be removed, but variations in terrain should also be corrected for. Multiple geometric transformations are applied to the image, depending on the perspective and terrain corrections required on a particular part of the image.

Orthophotos are commonly used in geographic information systems, such as are used by mapping agencies (e.g. Ordnance Survey) to create maps. Once the images have been aligned, or "registered", with known real-world coordinates, they can be widely deployed.

Large sets of orthophotos, typically derived from multiple sources and divided into "tiles" (each typically 256 x 256 pixels in size), are widely used in online map systems such as Google Maps. OpenStreetMap offers the use of similar orthophotos for deriving new map data. Google Earth overlays orthophotos or satellite imagery onto a digital elevation model to simulate 3D landscapes.

- Example

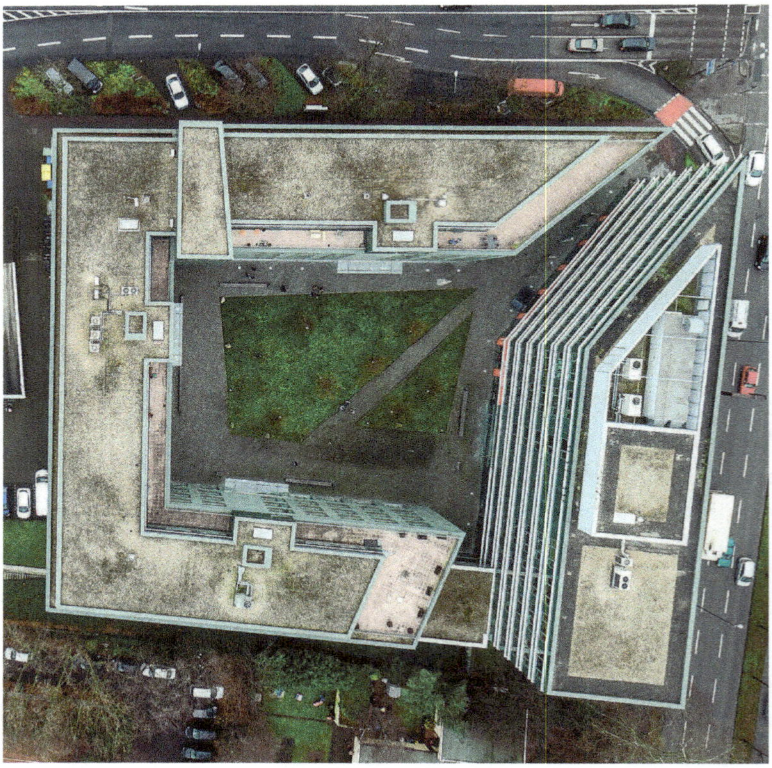

Aerial View Colonius Carre Köln, Germany

Atlantic City Aerial Photography

Aerial View Hotel Müggelsee in Berlin

Aerial photography Melaten cemetery Cologne

Aerial photography Hürth

Aerial View - Tennis club Read Green-White 1927,Germany

Aerial Video

With advancements in video technology, aerial video is becoming more popular. Orthogonal video is shot from aircraft mapping pipelines, crop fields, and other points of interest. Using GPS, video may be embedded with meta data and later synced with a video mapping program.

This "Spatial Multimedia" is the timely union of digital media including still photography, motion video, stereo, panoramic imagery sets, immersive media constructs, audio, and other data with location and date-time information from the GPS and other location designs.

Aerial videos are emerging Spatial Multimedia which can be used for scene understanding and object tracking. The input video is captured by low flying aerial platforms and typically consists of strong parallax from non-ground-plane structures. The integration of digital video, global positioning systems (GPS) and automated image processing will improve the accuracy and cost-effectiveness of data collection and reduction. Several different aerial platforms are under investigation for the data collection.

SATELLITE IMAGERY

Satellite imagery consists of images of Earth or other planets collected by satellites. Imaging satellites are operated by governments and businesses around the world. Satellite imaging companies sell images under license. Images are licensed to governments and businesses such as Apple Maps and Google Maps.

The first images from space were taken on the sub-orbital V-2 rocket flight launched by the U.S. on October 24, 1946.

HISTORY

The first images from space were taken on sub-orbital flights. The U.S-launched V-2 flight on October 24, 1946 took one image every 1.5 seconds. With an apogee of 65 miles (105 km), these photos were from five times higher than the previous record, the 13.7 miles (22 km) by the Explorer II balloon mission in 1935. The first satellite (orbital) photographs of Earth were made on August 14, 1959 by the U.S. Explorer 6. The first satellite photographs of the Moon might have been made on October 6, 1959 by the Soviet satellite Luna 3, on a mission to photograph the far side of the Moon.

The satellite images were made from pixels. The first crude image taken by the satellite Explorer 6 shows a sunlit area of the Central Pacific Ocean and its cloud cover. The photo was taken when the satellite was about 17,000 mi (27,000 km) above the surface of the earth on August 14, 1959. At the time, the satellite was crossing Mexico.

The Blue Marble photograph was taken from space in 1972, and has become very popular in the media and among the public. Also in 1972 the United States started the Landsat program, the largest program for acquisition of imagery of Earth from space. Landsat Data Continuity Mission, the most recent Landsat satellite, was launched on 11 February 2013. In 1977, the first real time satellite imagery was acquired by the United States's KH-11 satellite system.

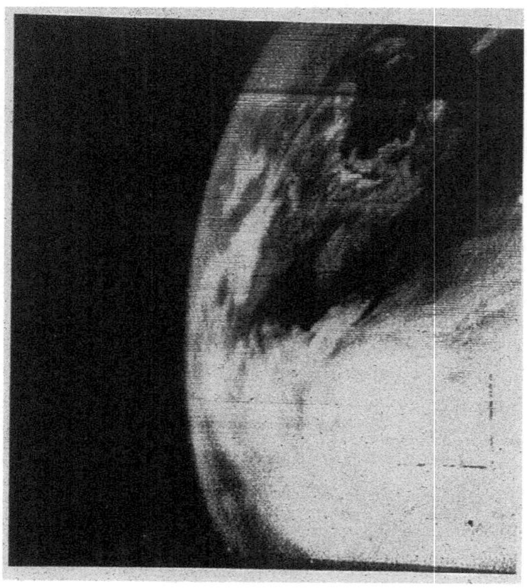

The first television image of Earth from space transmitted by the TIROS-1 weather satellite in 1960.

All satellite images produced by NASA are published by NASA Earth Observatory and are freely available to the public. Several other countries have satellite imaging programs, and a collaborative European effort launched the ERS and Envisat satellites carrying various sensors. There are also private companies that provide commercial satellite imagery. In the early 21st century satellite imagery became widely available when affordable, easy to use software with access to satellite imagery databases was offered by several companies and organizations.

Uses

Satellite images have many applications in meteorology, oceanography, fishing, agriculture, biodiversity conservation, forestry, landscape, geology, cartography, regional planning, education, intelligence and warfare. Images can be in visible colors and in other spectra. There are also elevation maps, usually made by radar images. Interpretation and analysis of satellite imagery is conducted using specialized remote sensing applications.

Satellite photography can be used to produce composite images of an entire hemisphere

...or to map a small area of the Earth, such as this photo of the countryside of Haskell County, Kansas, United States.

Resolution and Data

There are four types of resolution when discussing satellite imagery in remote sensing: spatial, spectral, temporal, and radiometric. Campbell (2002) defines these as follows:

- spatial resolution is defined as the pixel size of an image representing the size of the surface area (i.e. m^2) being measured on the ground, determined by the sensors' instantaneous field of view (IFOV);
- spectral resolution is defined by the wavelength interval size (discrete segment of the Electromagnetic Spectrum) and number of intervals that the sensor is measuring;
- temporal resolution is defined by the amount of time (e.g. days) that passes between imagery collection periods for a given surface location
- Radiometric resolution is defined as the ability of an imaging system to record many levels of brightness (contrast for example) and to the effective bit-depth of the sensor (number of grayscale levels) and is typically expressed as 8-bit (0-255), 11-bit (0-2047), 12-bit (0-4095) or 16-bit (0-65,535).
- Geometric resolution refers to the satellite sensor's ability to effectively image a portion of the Earth's surface in a single pixel and is typically expressed in

terms of Ground sample distance, or GSD. GSD is a term containing the overall optical and systemic noise sources and is useful for comparing how well one sensor can "see" an object on the ground within a single pixel. For example, the GSD of Landsat is ~30m, which means the smallest unit that maps to a single pixel within an image is ~30m x 30m. The latest commercial satellite (GeoEye 1) has a GSD of 0.41 m. This compares to a 0.3 m resolution obtained by some early military film based Reconnaissance satellite such as Corona.

The resolution of satellite images varies depending on the instrument used and the altitude of the satellite's orbit. For example, the Landsat archive offers repeated imagery at 30 meter resolution for the planet, but most of it has not been processed from the raw data. Landsat 7 has an average return period of 16 days. For many smaller areas, images with resolution as high as 41 cm can be available.

Satellite imagery is sometimes supplemented with aerial photography, which has higher resolution, but is more expensive per square meter. Satellite imagery can be combined with vector or raster data in a GIS provided that the imagery has been spatially rectified so that it will properly align with other data sets.

Imaging Satellites

GeoEye

GeoEye's GeoEye-1 satellite was launched September 6, 2008. The GeoEye-1 satellite has the high resolution imaging system and is able to collect images with a ground resolution of 0.41 meters (16 inches) in the panchromatic or black and white mode. It collects multispectral or color imagery at 1.65-meter resolution or about 64 inches.

DigitalGlobe

DigitalGlobe's WorldView-2 satellite provides high resolution commercial satellite imagery with 0.46 m spatial resolution (panchromatic only). The 0.46 meters resolution of WorldView-2's panchromatic images allows the satellite to distinguish between objects on the ground that are at least 46 cm apart. Similarly DigitalGlobe's QuickBird satellite provides 0.6 meter resolution (at NADIR) panchromatic images.

DigitalGlobe's WorldView-3 satellite provides high resolution commercial satellite imagery with 0.31 m spatial resolution. WVIII also carries a short wave infrared sensor and an atmospheric sensor

Spot Image

The 3 SPOT satellites in orbit (Spot 2, 4 and 5) provide images with a large choice of resolutions – from 2.5 m to 1 km. Spot Image also distributes multiresolution data from other optical satellites, in particular from Formosat-2 (Taiwan) and Kompsat-2 (South Korea) and from radar satellites (TerraSar-X, ERS, Envisat, Radarsat).

SPOT image of Bratislava

Spot Image will also be the exclusive distributor of data from the forthcoming very-high resolution Pleiades satellites with a resolution of 0.50 meter or about 20 inches. The first launch is planned for the end of 2011. The company also offers infrastructures for receiving and processing, as well as added value options.

ASTER

The Advanced Spaceborne Thermal Emission and Reflection Radiometer (ASTER) is an imaging instrument onboard Terra, the flagship satellite of NASA's Earth Observing System (EOS) launched in December 1999. ASTER is a cooperative effort between NASA, Japan's Ministry of Economy, Trade and Industry (METI), and Japan Space Systems (J-spacesystems). ASTER data is used to create detailed maps of land surface temperature, reflectance, and elevation. The coordinated system of EOS satellites, including Terra, is a major component of NASA's Science Mission Directorate and the Earth Science Division. The goal of NASA Earth Science is to develop a scientific understanding of the Earth as an integrated system, its response to change, and to better predict variability and trends in climate, weather, and natural hazards.

- Land surface climatology—investigation of land surface parameters, surface temperature, etc., to understand land-surface interaction and energy and moisture fluxes

- Vegetation and ecosystem dynamics—investigations of vegetation and soil distribution and their changes to estimate biological productivity, understand land-atmosphere interactions, and detect ecosystem change
- Volcano monitoring—monitoring of eruptions and precursor events, such as gas emissions, eruption plumes, development of lava lakes, eruptive history and eruptive potential
- Hazard monitoring—observation of the extent and effects of wildfires, flooding, coastal erosion, earthquake damage, and tsunami damage
- Hydrology—understanding global energy and hydrologic processes and their relationship to global change; included is evapotranspiration from plants
- Geology and soils—the detailed composition and geomorphologic mapping of surface soils and bedrocks to study land surface processes and earth's history
- Land surface and land cover change—monitoring desertification, deforestation, and urbanization; providing data for conservation managers to monitor protected areas, national parks, and wilderness areas

BlackBridge

BlackBridge, previously known as RapidEye, operates a constellation of five satellites, launched in August 2008, the RapidEye constillation contains identical multispectral sensors which are equally calibrated. Therefore, an image from one satellite will be equivalent to an image from any of the other four, allowing for a large amount of imagery to be collected (4 million km^2 per day), and daily revisit to an area. Each travel on the same orbital plane at 630 km, and deliver images in 5 meter pixel size. RapidEye satellite imagery is especially suited for agricultural, environmental, cartographic and disaster management applications. The company not only offers their imagery, but consults with their customers to create services and solutions based on analysis of this imagery .

ImageSat International

Earth Resource Observation Satellites, better known as "EROS" satellites, are lightweight, low earth orbiting, high-resolution satellites designed for fast maneuvering between imaging targets. In the commercial high-resolution satellite market, EROS is the smallest very high resolution satellite; it is very agile and thus enables very high performances. The satellites are deployed in a circular sun-synchronous near polar orbit at an altitude of 510 km (+/- 40 km). EROS satellites imagery applications are primarily for intelligence, homeland security and national development purposes but also employed in a wide range of civilian applications, including: mapping, border control, infrastructure planning, agricultural monitoring, environmental monitoring, disaster response, training and simulations, etc.

EROS A – a high resolution satellite with 1.9-1.2m resolution panchromatic was launched on December 5, 2000.

EROS B - the second generation of Very High Resolution satellites with 70 cm resolution panchromatic, was launched on April 25, 2006.

Meteosat

The Meteosat-2 geostationary weather satellite began operationally to supply imager data on 16 August 1981. Eumetsat has operated the Meteosats since 1987.

- The *Meteosat visible and infrared imager (MVIRI)*, three-channel imager: visible, infrared and water vapour; It operates on the first generation Meteosat, Meteosat-7 being still active.
- The 12-channel *Spinning Enhanced Visible and Infrared Imager (SEVIRI)* includes similar channels to those used by MVIRI, providing continuity in climate data over three decades; Meteosat Second Generation (MSG).
- The *Flexible Combined Imager (FCI)* on Meteosat Third Generation (MTG) will also include similar channels, meaning that all three generations will have provided over 60 years of climate data.

Model of a first generation Meteosat geostationary satellite.

Disadvantages

Because the total area of the land on Earth is so large and because resolution is relatively high, satellite databases are huge and image processing (creating useful images from the raw data) is time-consuming. Depending on the sensor used, weather conditions can affect image quality: for example, it is difficult to obtain images for areas of frequent cloud cover such as mountain-tops. For such reasons, publicly available satellite image datasets are typically processed for visual or scientific commercial use by third parties.

Commercial satellite companies do not place their imagery into the public domain and do not sell their imagery; instead, one must be licensed to use their imagery. Thus, the ability to legally make derivative products from commercial satellite imagery is minimized.

Privacy concerns have been brought up by some who wish not to have their property shown from above. Google Maps responds to such concerns in their FAQ with the following statement: "*We understand your privacy concerns... The images that Google Maps displays are no different from what can be seen by anyone who flies over or drives by a specific geographic location.*"

Moving Images

In 2005 the Australian company Astrovision (ASX: HZG) announced plans to launch the first commercial geostationary satellite in the Asia-Pacific. It is intended to provide true color, real-time live satellite feeds, with down to 250 meters resolution over the entire Asia-Pacific region, from India to Hawaii and Japan to Australia. They were going to provide this content to users of 3G mobile phones, over Pay TV as a weather channel, and to corporate and government users.

Potential customers were excited by the possibilities offered, but they were unwilling (or, in government cases, generally unable) to sign contracts for a service that would not be delivered for 3–4 years (the length of time required to build and launch the satellite). AstroVision ran low on funds and was forced to shut down the program in 2006.

REFERENCES

1. History of Aerial Photography Professional Aerial Photographers Association (retrieved 5 October 2016)
2. Ron Graham and Roger E. Read, Manual of Aerial Photography, London and Boston, Focal Press, ISBN 0-240-51229-4
3. Staff writer (April 3, 2013). "This Picture of Boston, Circa 1860, Is the World's Oldest Surviving Aerial Photo". Smithsonian Magazine. Retrieved April 17, 2013.
4. Archibald, Douglas (1897). "The Story of the Earth's Atmosphere". p. 174. Retrieved 2011-04-16.
5. Benton, Cris (June 25, 2010). "The First Kite Photographs". arch.ced.berkeley.edu. Archived from the original on 2011-06-09. Retrieved 2011-04-16.
6. "Arthur Batut Museum" (in French). Retrieved 2008-01-08.
7. "A Brief History of Aerial Photography".
8. "Royal Flying Corps Founded". History Today.
9. Marshall Cavendish Corporation (2003). How It Works: Science and Technology. Marshall Cavendish. p. 33.
10. "Lieutenant Leonard T.E. Taplin, D.F.C". Southsearepublic.org. Archived from the original on 2012-07-15. Retrieved 2013-01-24.
11. BFI Screenonline: Friese-Greene, Claude (1898-1943) Biography
12. "Municipal Air Surveys. Contracts From Doncaster And Birkenhead". The Times. issue:44229,column:E. Gale:The Times digital archive 1785–1985. 25 March 1926. p. 11. Retrieved 30 August 2012. (subscription required)

13. The Complete Encyclopedia of World Aircraft. 1997. p. 382. ISBN 0-7607-0592-5.
14. "Wide Area Is Mapped From Air By Giant Ten Lens Camera" Popular Mechanics, October 1935 editors have stated Fairchild Aircraft in hand written comment to left of archived article
15. "Nine Lens Aerial Camera Films 600 Square Miles" ''Popular Mechanics", April 1936. Books.google.com. Retrieved 2013-01-24.
16. "Photography Before Edgerton".
17. Cotton, Sidney (1969). Aviator Extraordinary: The Sidney Cotton Story. Chatto & Windus. p. 169. ISBN 0-7011-1334-0.
18. "Aerial Photography and Remote Sensing". University of Colorado Boulder. 2011. Retrieved March 25, 2011.
19. Mills, J.P.; et al. (1997). "Photogrammetry from Archived Digital Imagery for Seal Monitoring". The Photogrammetric Record. 15 (89): 715–724. doi:10.1111/0031-868X.00080.
20. Twiss, S.D.; et al. (2001). "Topographic spatial characterisation of grey seal Halichoerus grypus breeding habitat at a sub-seal size spatial grain". Ecography. 24 (3): 257–266. doi:10.1111/j.1600-0587.2001.tb00198.x.
21. Stewart, J.E.; et al. (2014). "Finescale ecological niche modeling provides evidence that lactating gray seals (Halichoerus grypus) prefer access to fresh water in order to drink". Marine Mammal Science. 30 (4): 1456–1472. doi:10.1111/mms.12126.
22. Yan G.; et al.(2007). "Automatic extraction of power lines from aerial images", IEEE Geoscience and Remote Sensing Letters, 4(3):387-391.
23. Stark, M. T., Griffin, P., Phoeurn, C., Ledgerwood, J., Dega, M., Mortland, C., ... & Latinis, K. (1999). Results of the 1995-1996 archaeological field investigations at Angkor Borei, Cambodia. Asian Perspectives 38(1)
24. United States Federal Aviation Regulations FAR part 91 section 119(14CFR91.119)
25. http://www.gpo.gov/fdsys/pkg/FR-2014-06-25/pdf/2014-14948.pdf
26. "Civil Aviation Safety Regulations 1998". Australian Civil Aviation Safety Authority. 2002-12-19. Retrieved 2013-01-24.
27. "Civil Aviation Safety Authority". Australian Civil Aviation Safety Authority. 2002-12-19. Retrieved 2013-01-24.
28. "List of UAS Operator Certificate Holders". Australian Civil Aviation Safety Authority. Retrieved 2013-01-24.
29. "Part 101 Amendments - Cutting red tape for remotely piloted aircraft". CASA. 2016. Retrieved 5 October 2016.
30. Huerta, Michael P. (18 June 2014). "Interpretation of the Special Rule for Model Aircraft" (PDF). FAA. Retrieved 5 October 2016.
31. "U.S. Transportation Secretary Foxx Announces FAA Exemptions for Commercial UAS Movie and TV Production". Press Release. FAA. 25 September 2014. Retrieved 4 October 2014.

32. Public Law 112–95, section 336(c).
33. California v. Ciraolo, 476 U.S. 206 (1986)
34. "Lecture 6.1: Classification of Photographs". The Remote Sensing Core Curriculum. University of Maryland, Baltimore County. 1999. Retrieved 2011-03-25.
35. Short, Nicholas (2010-04-28). "Elements of Aerial Photography". Remote Sensing Tutorial Page 10-1. NASA. Archived from the original on March 17, 2011. Retrieved 2011-03-25.

CHAPTER 8

Remote Sensing Application

Chapter Contents

- **Introduction**
- **PCI Geomatica**
- **ERDAS Imagine**
- **Exelis Visual Information Solutions**
- **Google Earth**
- **Grass GIS**
- **Remote view**
- **Cognition Network Technology**

INTRODUCTION

A remote sensing application is a software application that processes remote sensing data. Remote sensing applications are similar to graphics software, but they enable generating geographic information from satellite and airborne sensor data. Remote sensing applications read specialized file formats that contain sensor image data, georeferencing information, and sensor metadata. Some of the more popular remote sensing file formats include: GeoTIFF, NITF, JPEG 2000, ECW (file format), MrSID, HDF, and NetCDF.

Remote Sensing applications perform many features including:

- Change Detection — Determine the changes from images taken at different times of the same area
- Orthorectification — Warp an image to its location on the earth.
- Spectral Analysis — For example, using non-visible parts of the electromagnetic spectrum to determine if a forest is healthy
- Image Classification — Categorization of pixels based on reflectance into different land cover classes (e.g. Supervised classification, Unsupervised classification and Object Oriented Classification)

Many remote sensing applications are built using common remote sensing toolkits, like GDAL and OSSIM.

Examples of Remote Sensing Software

- Geomatica, PCI Geomatics
- TNTmips gis, MicroImages,USA
- ERDAS IMAGINE
- ENVI
- Google Earth
- GRASS GIS
- OpenEV (Open Source)
- Opticks (Open Source)
- Orfeo toolbox (Open Source)
- RemoteView
- SOCET SET
- IDRISI
- ECognition

PCI GEOMATICA

PCI Geomatica is a remote sensing desktop software package for processing earth observation data, designed by PCI Geomatics. The latest version of the software is Geomatica 2014. Geomatica is aimed primarily at faster data processing and allows users to load satellite and aerial imagery where advanced analysis can be performed. Geomatica has been used by many educational institutions and scientific programs throughout the world to analyze satellite imagery and trends, such as the GlobeSAR Program, a program which was carried out by the Canada Centre for Remote Sensing in the 1990s.

A very popular edition of Geomatica is known as Freeview, which permits users to load multiple types of satellite images as well as geospatial data that is stored in different formats. The software is available for download over the web, and has registered several thousands of downloads. CNET Download page for Freeview.

Image Processing Packages

Geomatica is one of several software packages available to the educational, commercial, and military users. Other similar packages include Erdas Imagine, Envi, and SocetSet (or Socet GXP). An independent review of the software and its functionality written by Directions Magazine is included here: http://www.directionsmag.com/articles/product-review-pci146s-geomatica-10/123136. Geomatica has also been compared to Envi and Erdas Imagine as it relates to orthorectification. http://www.isprs.org/proceedings/XXXVII/congress/4_pdf/283.pdf

Educational Institutions using Geomatica

Over 2,700 educational institutions worldwide have used Geomatica as part of their Remote Sensing course delivery, some of which are listed here

- University of Calgary, Geomatics Engineering program
- University of New Brunswick, Online Course offered on Radarsat-2 and Polarimetry
- Fleming College, Lindsay, Ontario, Canada
- Carleton University, Ottawa, Canada
- University of Waterloo, Waterloo, Ontario, Canada
- British Columbia Institute of Technology (BCIT)
- Université du Québec à Montréal, Remote Sensing course GEO8142
- Aalto University, Institute of Photogrammetry and Remote Sensing
- North Eastern University (NEU), Boston, USA
- University of Arkansas
- University of Victoria
- Fanshawe College, London, Ontario
- TU Bergakademie Freiberg, Remote Sensing Group
- Saint Mary›s University (SMU), Halifax, Nova Scotia, Canada. Department of Geography and Environmental Studies

Open Geospatial Consortium

Geomatica includes a web coverage service interface that complies with the OGC Web Coverage Service (WCS) Interface Standard, which is a key area in which PCI Geomatics has contributed. Remote Sensing data providers distribute data in diverse formats, which makes sharing information across many different platforms challenging. WCS seeks to alleviate some of the data sharing challenges by publishing the geographic information and layers openly over the web.

TNTmips

TNTmips is a geospatial analysis system providing a fully featured GIS, RDBMS, and automated image processing system with CAD, TIN, surface modeling, map layout and innovative data publishing tools. TNTmips has a single integrated system with an identical interface, functionality, and geodata structure for use on Mac and Windows operating systems. The interface, database text content, messages, map production, and all other internal aspects of TNTmips have been localized for use in many languages, including, for example Arabic, Thai, and all romance languages. The professional version of TNTmips is in use in over 120 nations while the TNTmips Free version (restricted in project size) is used worldwide for educational, self learning, and small projects (e.g., archaeological sites, neighborhood planning, and precision farming).

General Information

TNTmips is a professional system for fully integrated GIS, image processing, CAD, TIN, desktop cartography, and geospatial database management.

License Levels

TNTmips Pro (professional license, TNTmips Basic (low cost license), TNTmips Free (freeware).

Special Academic License (SAL)

The TNT products include: TNTmips, TNTedit, TNTview, TNTatlas and TNTsdk. There is no distinction between TNTmips and TNT products with regards to license levels. All TNT products are available for Windows and Macintosh computers in a growing number of international languages.

ERDAS IMAGINE

ERDAS Imagine is a remote sensing application with raster graphics editor abilities designed by ERDAS for geospatial applications. The latest version is 2015. Imagine is aimed mainly at geospatial raster data processing and allows users to prepare, display and enhance digital images for mapping use in geographic information system (GIS) and computer-aided design (CAD) software. It is a toolbox allowing the user to perform numerous operations on an image and generate an answer to specific geographical questions.

By manipulating imagery data values and positions, it is possible to see features that would not normally be visible and to locate geo-positions of features that would otherwise be graphical. The level of brightness, or reflectance of light from the surfaces in the image can be helpful with vegetation analysis, prospecting for minerals etc. Other usage examples include linear feature extraction, generation of processing work flows (*spatial models* in Imagine), import/export of data for a wide variety of formats, orthorectification, mosaicking of imagery, stereo and automatic feature extraction of map data from imagery.

Product History

Before the ERDAS Imagine Suite, ERDAS, Inc. developed various products to process satellite imagery from AVHRR, Landsat MSS and TM, and Spot Image into land cover, land use maps, map deforestation, and assist in locating oil reserves under the product name ERDAS. These older ERDAS applications were rewritten from Fortran to C and C++ and exist today within the Imagine Suite, which has grown to support most optical and radar mapping satellites, airborne mapping cameras and digital sensors used for mapping.

ERDAS 4

The first version of ERDAS was launched in 1978 on Cromemco microcomputers based on the 8-bit Z80 CPU running the CDOS operating system. The system was built into a desk and was configured with one color monitor (256 x 256 resolution), one B&W monitor, two 8" floppy drives (one for software and one for data). Later, other options were added, such as a large digitizing tablet, and a hard drive (hard drives usually did not exist on computers this size). The hard drive was a Control Data Corporation model about the size of a small washing machine. It had 80 MB fixed disks and a very large 16 MB removable platter. When the hard disk was installed, ERDAS engineers had to cut out the back of the computer furniture to make room for the cables

ERDAS 400

In 1980, ERDAS, Inc. developed the ERDAS 400. This early system was a turnkey computerized forest-management system for the state of New York. The hardware needed to make this ability available at an affordable price did not exist at the time, therefore some circuit board and other hardware designs were created by ERDAS to deliver the product. Between 1980 and 1982 different versions of ERDAS 400 legacy were delivered to NASA, US Forest Service, US Environmental Protection Agency, the State of Illinois, and others.

ERDAS 7.x

In November 1982 ERDAS 7.0 was released. This product was a change from the more turnkey, custom design of the ERDAS 400 systems to a DOS IBM Personal Computer. These systems continued to be command prompt based, but added a nested menu assisting customers through prompts needed to process the imagery into data used in a GIS.

In ERDAS 7.2 (Jan 1983) the first PC version of ERDAS software was licensed to the University of South Carolina Department of Geography. The release of versions ERDAS 7.0 - 7.5 marked a period of wide acceptance of remote sensing technology among US federal and state government agencies, and research universities needing to monitor natural resource changes.

A link between ESRI ARC/INFO and ERDAS 7.3 was introduced in 1988. The "ERDAS-ARC/INFO Live Link" allowed the mapping community to use technology from both companies to deliver high quality image display and image processing with powerful GIS ability. This relationship between ESRI and ERDAS, Inc. remained strong at a high management level and in various product offerings into 2007. This strong relationship began very early, as the ERDAS founders Lawrie Jordan and Bruce Rado and ESRI founder Jack Dangermond all attended the Harvard Graduate School of Design.

The final version of the ERDAS product line was ERDAS 7.5, which was finished as the new product, ERDAS Imagine, was being developed. All the abilities of ERDAS 7.5 were merged into Imagine.

ERDAS Imagine

ERDAS Imagine was demonstrated in October 1991 and released as ERDAS Imagine 8.0 in February 1992. It was released on a Sun Workstation using SunOS providing a graphical user interface (GUI) to assist in visualizing imagery used in mapping, vector GIS data, creating maps, and so forth. Much of the ERDAS 7.5 product was provided for several years through intermediate product releases of ERDAS 8.01 & 8.02 until the ERDAS Imagine product replaced all the ERDAS 7.5 ability in 1994 with the release of Imagine 8.1.

The relationship between the ERDAS and ARC/INFO products began with the "ERDAS-ARC/Info Live Link" and was expanded in ERDAS Imagine 8.2 (Feb 1993), when ERDAS released the Imagine Vector Module. This add-on was fully developed by ERDAS, but licensed the Arc Coverage data format from ESRI. The vector module is one of the most popular modules within the Imagine product suite through the 9.3 release (Sept 2008).

The processing of radar data in ERDAS Imagine began with the development of the Radar Module, first released in Nov. 1992. The abilities have been upgraded many times since with the original tools now named *Radar Interpreter*.

ERDAS Imagine Spatial Modeler

With the release of ERDAS 7.5 in 1990 ERDAS introduced a GIS Modeling (GISMO) script language. Tools were made available to assist in the development of complex spatial models using Dana Tomlin's map algebra concepts. GISMO was upgraded in 1992 with the release of ERDAS Imagine 8.0 to the Spatial Modeler scripting language. Model Maker, a graphic flow chart model building enhancement to Spatial Modeler was introduced in 1993. In 2004 ESRI copied the graphic flow chart model building environment idea and released an ArcGIS component named Model Builder.

With the release of ERDAS Imagine 2013 in December 2012, the Model Maker's graphical user interface was changed and its architecture was rewritten from a push model to a pull model. This re-architecture allows many complex models to be processed directly to a viewer in near-real time. A Python capability was added at this time.

A variety of ready to go spatial models can be downloaded from Sterling Geo's website who distribute the software in the UK.

EXELIS VISUAL INFORMATION SOLUTIONS

Exelis Visual Information Solutions (Exelis VIS), a wholly owned subsidiary of ITT Exelis, provides software for the analysis and visualization of scientific data and imagery. The company develops products such as IDL, ENVI and IAS which are used in a variety of industries, including remote sensing, engineering, geology, aerospace, defense and intelligence, and medical imaging.

History

When a group of scientists who developed the Mariner Mars space probes in 1977, David Stern, who was working at the University of Colorado developed a programming language called Mariner Mars Spectral Editor, which later evolved to IDL, Interactive Data Language, in order to supplement FORTRAN, specifically for data analysis and visualization at that time. As his vision developed, Stern left the university and started working in his attic to improve the programming language. He incorporated Research Systems Inc. (RSI) and released IDL as a proprietary programming language for visualizing data. NASA Goddard Space Center and Ball Aerospace were ones of the company's early customers.

In 1994, the software package, ENVI, was released. Written in IDL, ENVI enhanced the text-based IDL and enabled a suite of graphical user interfaces specialized in remote sensing imagery analysis. RSI was acquired by Eastman Kodak as a wholly owned subsidiary in 2000. Four years later, RSI along with the Remote Sensing Systems division of Eastman Kodak, were sold to ITT Corporation and were renamed ITT Visual Information Solutions.

In 2011, ITT Corporation was divided into three standalone companies, and ITT Visual Information Solutions became part of Exelis Inc. along with ITT Defense and Information Solutions. ITT Visual Information Solutions was renamed again to the current Exelis Visual Information Solutions.

In 2015 Exelis Visual Information Solutions was purchased by the Harris Corporation, becoming Harris Geospatial.

Products

IDL

Main article: IDL (programming language)

IDL (Interactive Data Language), is a programming language used for data analysis. It is used in particular areas of science, such as astronomy and medical imaging.

ENVI

ENVI (an acronym for "ENvironment for Visualizing Images") is a software application used to process and analyze geospatial imagery. It is commonly used by remote sensing professionals and image analysts.

ENVI bundles together a number of scientific algorithms for image processing a lot of which are contained in automated, wizard-based approach that walks users through complex tasks.It was originally developed by Better Solutions Consulting, LLC, a partnership of five individuals in Boulder, CO. In 2000, ENVI was sold to Kodak at the same time Kodak purchased RSI.

GOOGLE EARTH

Google Earth is a virtual globe, map and geographical information program that was originally called EarthViewer 3D created by Keyhole, Inc, a Central Intelligence Agency (CIA) funded company acquired by Google in 2004. It maps the Earth by the superimposition of images obtained from satellite imagery, aerial photography and geographic information system (GIS) onto a 3D globe. It was originally available with three different licenses, but has since been reduced to just two: Google Earth (a free version with limited function) and Google Earth Pro, which is now free (it previously cost $399 a year) and is intended for commercial use. The third original option, Google Earth Plus, has been discontinued.

The product, re-released as Google Earth in 2005, is available for use on personal computers running Windows 2000 and above, Mac OS X 10.3.9 and above, Linux kernel: 2.6 or later (released on June 12, 2006), and FreeBSD. Google Earth is also available as a browser plugin which was released on May 28, 2008. It was also made available for mobile viewers on the iPhone OS on October 28, 2008, as a free download from the App Store, and is available to Android users as a free app in the Google Play store. In addition to releasing an updated Keyhole based client, Google also added the imagery from the Earth database to their web-based mapping software, Google Maps. The release of Google Earth in June 2005 to the public caused a more than tenfold increase in media coverage on virtual globes between 2004 and 2005, driving public interest in geospatial technologies and applications. As of October 2011, Google Earth has been downloaded more than a billion times.

Google Earth displays satellite images of varying resolution of the Earth's surface, allowing users to see things like cities and houses looking perpendicularly down or at an oblique angle. The degree of resolution available is based somewhat on the points of interest and popularity, but most land (except for some islands) is covered in at least 15 meters of resolution. Maps showing a visual representation of Google Earth coverage Melbourne, Victoria, Australia; Las Vegas, Nevada, United States; and Cambridge, Cambridgeshire, United Kingdom include examples of the highest resolution, at 15 cm (6 inches). Google Earth allows users to search for addresses for some countries, enter coordinates, or simply use the mouse to browse to a location.

For large parts of the surface of the Earth only 2D images are available, from almost vertical photography. Viewing this from an oblique angle, there is perspective in the sense that objects which are horizontally far away are seen smaller, like viewing a large photograph, not quite like a 3D view.

For other parts of the surface of the Earth, 3D images of terrain and buildings are available. Google Earth uses digital elevation model (DEM) data collected by NASA's Shuttle Radar Topography Mission (SRTM). This means one can view almost the entire earth in three dimensions. Since November 2006, the 3D views of many mountains, including Mount Everest, have been improved by the use of supplementary DEM data to fill the gaps in SRTM coverage.

Some people use the applications to add their own data, making them available through various sources, such as the Bulletin Board Systems (BBS) or blogs mentioned in the link section below. Google Earth is able to show various kinds of images overlaid on the surface of the earth and is also a Web Map Service client. Google Earth supports managing three-dimensional Geospatial data through Keyhole Markup Language (KML).

Detail

Google Earth is simply based on 3D maps, with the capability to show 3D buildings and structures (such as bridges), which consist of users' submissions using SketchUp, a 3D modeling program software. In prior versions of Google Earth (before Version 4), 3D buildings were limited to a few cities, and had poorer rendering with no textures. Many buildings and structures from around the world now have detailed 3D structures; including (but not limited to) those in the United States, Canada, Mexico, India, Japan, United Kingdom, Spain, Germany, Pakistan and the cities, Amsterdam and Alexandria. In August 2007, Hamburg became the first city entirely shown in 3D, including textures such as façades. The 'Westport3D' model was created by 3D imaging firm AM3TD using long-distance laser scanning technology and digital photography and is the first such model of an Irish town to be created. As it was developed initially to aid Local Government in carrying out their town planning functions it includes the highest-resolution photo-realistic textures to be found anywhere in Google Earth. Three-dimensional renderings are available for certain buildings and structures around the world via Google's 3D Warehouse and other websites. In June 2012, Google announced that it will start to replace user submitted 3D buildings with auto-generated 3D mesh buildings starting with major cities. Although there are many cities on Google Earth that are fully or partially 3D, more are available in the Earth Gallery. The Earth Gallery is a library of modifications of Google Earth people have made. In the library there are not only modifications for 3D buildings, but also models of earthquakes using the Google Earth model, 3D forests, and much more.

Recently, Google added a feature that allows users to monitor traffic speeds at loops located every 200 yards in real-time. In 2007, Google began offering traffic data in real-time, based on information crowdsourced from the GPS-identified locations of cellular phone users. In version 4.3 released on April 15, 2008, Google Street View was fully integrated into the program allowing the program to provide an on the street level view in many locations.

On January 31, 2010, the entirety of Google Earth's ocean floor imagery was updated to new images by SIO, NOAA, US Navy, NGA, and GEBCO. The new images have caused smaller islands, such as some atolls in the Maldives, to be rendered invisible despite their shores being completely outlined.

Uses

Google Earth may be used to perform some day-to-day tasks and for other purposes.

- Google Earth can be used to view areas subjected to widespread disasters if Google supplies up-to-date images. For example, after the January 12, 2010 Haiti earthquake images of Haiti were made available on January 17.
- With Google's push for the inclusion of Google Earth in the Classroom, teachers are adopting Google Earth in the classroom for lesson planning, such

as teaching students geographical themes (location, culture, characteristics, human interaction, and movement) to creating mashups with other web applications such as Wikipedia.

- One can explore and place location bookmarks on the Moon and Mars.
- One can get directions using Google Earth, using variables such as street names, cities, and establishments. But the addresses must by typed in search field, one cannot simply click on two spots on the map.
- Google Earth can function as a hub of knowledge, pertaining the users location. By enabling certain options, one can see the location of gas stations, restaurants, museums, and other public establishments in their area. Google Earth can also dot the map with links to images, YouTube videos, and Wikipedia articles relevant to the area being viewed.
- One can create custom image overlays for planning trips, hikes on handheld GPS units.
- Google Earth can be used to map homes and select a random sample for research in developing countries.

All of these features are also released by Google Earth Blog.

Features

Wikipedia and Panoramio integration

In December 2006, Google Earth added a new layer called "Geographic Web" that includes integration with Wikipedia and Panoramio. In Wikipedia, entries are scraped for coordinates via the Coord templates. There is also a community-layer from the project Wikipedia-World. More coordinates are used, different types are in the display and different languages are supported than the built-in Wikipedia layer. Google announced on May 30, 2007 that it is acquiring Panoramio. In March 2010, Google removed the "Geographic Web" layer. The "Panoramio" layer became part of the main layers and the "Wikipedia" layer was placed in the "More" layer.

Flight Simulator

In Google Earth v4.2 a flight simulator was included as a hidden feature. Starting with v4.3 it is no longer hidden. The flight simulator could be accessed by holding down the keys Ctrl, Alt, and A. Initially the F-16 Fighting Falcon and the Cirrus SR-22 were the only aircraft available, and they could be used with only a few airports. However, one can start flight in "current location" and need not to be at an airport. One will face the direction they face when they start the flight simulator.

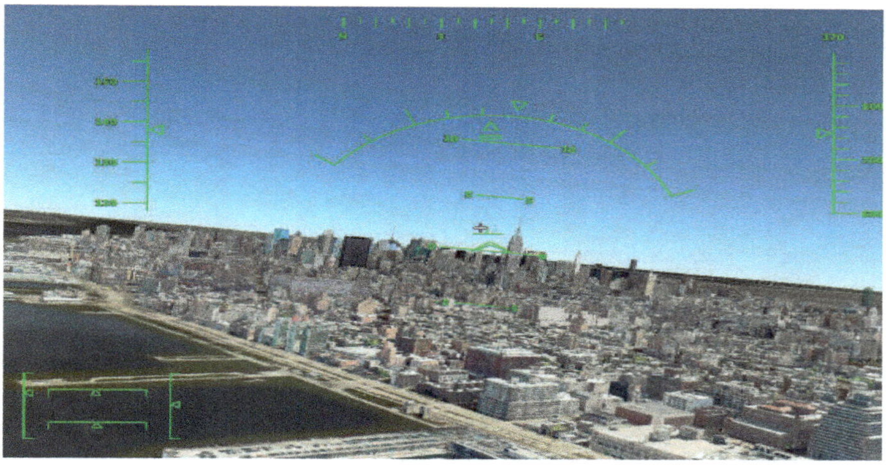

Downtown Toronto as seen from a F-16 Fighting Falcon during a simulated flight

They cannot start flight in ground level view and must be near the ground (approximately 50m-100m above the ground) to start in take-off position. Otherwise they will be in the air with 40% flaps and gears extended (landing position). In addition to keyboard control, the simulator can be controlled with a mouse or joystick. Google Earth v5.1 and higher crashes when starting flight simulator with Saitek and other joysticks. The user can also fly underwater.

Featured planes

- F-16 Fighting Falcon – A much higher speed and maximum altitude than the Cirrus SR-22, it has the ability to fly at a maximum speed of Mach 2, although a maximum speed of 1678 knots (3108 km/h) can be achieved. The take-off speed is 225 knots, the landing speed is 200 knots (370 km/h).
- Cirrus SR-22 – Although slower and with a lower maximum altitude, the SR-22 is much easier to handle and is preferred for up-close viewing of Google Earth's imagery. The take-off speed is 75 knots (139 km/h), the landing speed is 70 knots (130 km/h)

The flight simulator can be commanded with the keyboard, mouse or plugged-in joystick. Broadband connection and a high speed computer provides a very realistic experience. The simulator also runs with animation, allowing objects (for example: planes) to animate while on the simulator. Programming language can also be used to make it look like the cockpit of a plane, or for instrument landing.

Sky Mode

Google Sky is a feature that was introduced in Google Earth 4.2 on August 22, 2007, and allows users to view stars and other celestial bodies. It was produced by Google through a partnership with the Space Telescope Science Institute (STScI) in Baltimore, the science operations center for the Hubble Space Telescope. Dr. Alberto Conti and his co-developer Dr. Carol Christian of STScI plan to add the public images from 2007, as well as color images of all of the archived data from Hubble's Advanced Camera for Surveys. Newly released Hubble pictures will be added to the Google Sky program as soon as they are issued. New features such as multi-wavelength data, positions of major satellites and their orbits as well as educational resources will be provided to the Google Earth community and also through Christian and Conti's website for Sky. Also visible on Sky mode are constellations, stars, galaxies and animations depicting the planets in their orbits. A real-time Google Sky mashup of recent astronomical transients, using the VOEvent protocol, is being provided by the VOEventNet collaboration. Google's Earth maps are being updated each 5 minutes.

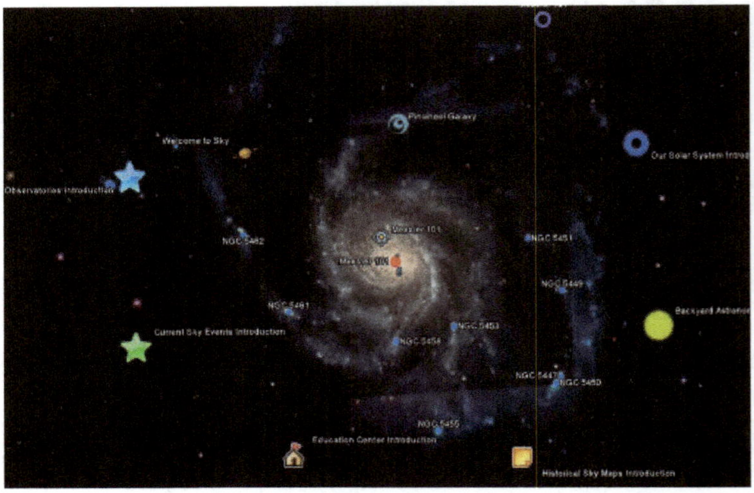

Google Earth in Sky Viewing Mode

Google Sky faces competition from Microsoft WorldWide Telescope (which runs only under the Microsoft Windows operating systems) and from Stellarium, a free open source planetarium that runs under Microsoft Windows, OS X, and Linux.

On March 13, 2008, Google made a web-based version of Google Sky available via the internet.

Street View

On April 15, 2008 with version 4.3, Google fully integrated its Street View into Google Earth. In version 6.0, the photo zooming function has been removed because it is incompatible with the new 'seamless' navigation.

Google Street View provides 360° panoramic street-level views and allows users to view parts of selected cities and their surrounding metropolitan areas at ground level. When it was launched on May 25, 2007 for Google Maps, only five cities were included. It has since expanded to more than 40 U.S. cities, and includes the suburbs of many, and in some cases, other nearby cities. Recent updates have now implemented Street View in most of the major cities of Canada, Mexico, Denmark, South Africa, Japan, Spain, Norway, Finland, Sweden, France, the UK, Republic of Ireland, the Netherlands, Italy, Switzerland, Portugal, Taiwan, and Singapore.

Google Street View, when operated, displays photos that were previously taken by a camera mounted on an automobile, and can be navigated by using the mouse to click on photograph icons displayed on the screen in the user's direction of travel. Using these devices, the photos can be viewed in different sizes, from any direction, and from a variety of angles.

Water and Ocean

Introduced in version 5.0 (February 2009), the *Google Ocean* feature allows users to zoom below the surface of the ocean and view the 3D bathymetry beneath the waves. Supporting over 20 content layers, it contains information from leading scientists and oceanographers. On April 14, 2009, Google added underwater terrain data for the Great Lakes. In 2010, Google added underwater terrain data for Lake Baikal.

In June 2011, higher resolution of some deep ocean floor areas increased in focus from 1-kilometer grids to 100 meters thanks to a new synthesis of seafloor topography released through Google Earth. The high-resolution features were developed by oceanographers at Columbia University's Lamont-Doherty Earth Observatory from scientific data collected on research cruises. The sharper focus is available for about 5 percent of the oceans (an area larger than North America). Underwater scenery can be seen of the Hudson Canyon off New York City, the Wini Seamount near Hawaii, and the sharp-edged 10,000-foot-high Mendocino Ridge off the U.S Pacific Coast. There is a Google 2011 Seafloor Tour for those interested in viewing ocean deep terrain.

Historical Imagery

Introduced in version 5.0, Historical Imagery allows users to traverse back in time and study earlier stages of any place. This feature allows research that require analysis of past records of various places.

Mars

A picture of Mars' landscape.

Google Earth 5 includes a separate globe of the planet Mars, that can be viewed and analysed for research purposes. The maps are of a much higher resolution than those on the browser version of Google Mars and it also includes 3D renderings of the Martian terrain. There are also some extremely-high-resolution images from the Mars Reconnaissance Orbiter's HiRISE camera that are of a similar resolution to those of the cities on Earth. Finally, there are many high-resolution panoramic images from various Mars landers, such as the Mars Exploration Rovers, Spirit and Opportunity, that can be viewed in a similar way to Google Street View. Interestingly enough, layers on Google Earth (such as World Population Density) can also be applied to Mars. Layers of Mars can also be applied onto Earth. Mars also has a small application found near the face on Mars. It is called Meliza, and features a chat between the user and an automatic robot speaker. It is useful for research on Mars, but is not recommended for normal conversations.

Moon

One of the lunar landers viewed in Google Moon

On July 20, 2009, the 40th anniversary of the Apollo 11 mission, Google introduced the Google Earth version of Google Moon, which allows users to view satellite images of the Moon. It was announced and demonstrated to a group of invited guests by Google along with Buzz Aldrin at the Newseum in Washington, D.C.

Google Earth Engine

Google Earth Engine is a separate product, not a part of Google Earth.

Liquid Galaxy

Liquid Galaxy is a cluster of computers running Google Earth creating an immersive experience. On September 30, 2010, Google made the configuration and schematics for their rigs public, placing code and setup guides on the Liquid Galaxy wiki.

Liquid Galaxy has also been used as a panoramic photo viewer using KRpano, as well as a Google Street View viewer using Peruse-a-Rue Peruse-a-Rue is a method for synchronizing multiple Maps API clients.

Influences

Google Earth can be traced directly back to a small company named Autometric, now a part of Boeing. A team at Autometric, led by Robert Cowling, created a visualization product named Edge Whole Earth. Bob demonstrated Edge to Michael T. Jones, Chris Tanner and others at SGC in 1996. Several other visualization products using imagery existed at the time, including Performer-based ones, but Michael T. Jones stated emphatically that he had "never thought of the complexities of rendering an entire globe ..." The catch phrase "from outer space to in your face" was coined by Autometric President Dan Gordon, and used to explain his concept for personal/local/global range. Edge blazed a trail as well in broadcasting, being used in 1997 on CBS News with Dan Rather, in print for rendering large images draped over terrain for National Geographic, and used for special effects in the feature film *Shadow Conspiracy* in 1997.

Gordon was a huge fan of the 'Earth' program described in Neal Stephenson's sci-fi classic *Snow Crash*. Indeed, a Google Earth co-founder claimed that Google Earth was modeled after *Snow Crash*, while another co-founder said it was inspired by the short science education film *Powers of Ten*. In fact Google Earth was at least partly inspired by a Silicon Graphics demo called "From Outer Space to in Your Face" which zoomed from space into the Swiss Alps then into the Matterhorn. This launch demo was hosted by an Onyx 3000 with InfiniteReality4 graphics, which supported Clip Mapping and was inspired by the hardware texture paging capability (although it did not use the Clip Mapping) and "Powers of Ten". The first Google Earth implementation called Earth Viewer emerged from Intrinsic Graphics as a demonstration of Chris Tanner's software based implementation of a Clip Mapping texture paging system and was spun off as Keyhole Inc.

Versions and variations

Mac Version

Since version 4.1.7076.4558 (released on May 9, 2007) onward OS X users can, among other new features, upgrade to the "Plus" version via an option in the Google Earth menu. Some users reported difficulties with Google Earth crashing in the then current version when zooming in. Version 5 of Google Earth for Mac was released in 2009, and version 7 was released concomitantly with the Mac and PC versions on 31 October 2012.

Linux Version

Starting with the version 4 beta Google Earth functions under Linux, as a native port using the Qt toolkit. The Free Software Foundation consider the development of a free compatible client for Google Earth to be a High Priority Free Software Project.

Android Version

Google Earth running on Android

An Android version was released on Monday, February 22, 2010.

iOS Version

A version for the iOS, which runs on the iPhone, iPod Touch and the iPad, was released for free on the App Store on October 27, 2008.It makes use of the multi-touch interface to move on the globe, zoom or rotate the view, and allow to select the current location using the iPhone integrated Assisted GPS. Although it previously did not support any layers apart from Wikipedia and Panoramio, version 6.2 brought KML support to add additional layers. Version 7 introduced 3D modeling of several cities.

Google Earth Plus (Discontinued in 2008)

Discontinued in December 2008, Google Earth Plus was an individual-oriented paid subscription upgrade to Google Earth that provided customers with the following features, most of which are now available in the free Google Earth.

- GPS integration: read tracks and waypoints from a GPS device. A variety of third-party applications have been created which provide this functionality using the basic version of Google Earth by generating KML or KMZ files based on user-specified or user-recorded waypoints. However, Google Earth Plus provides direct support for the Magellan and Garmin product lines, which together hold a large share of the GPS market. The Linux version of the Google Earth Plus application does not include any GPS functionality.
- Higher-resolution printing.
- Customer support via email.
- Data importer: read address points from CSV files; limited to 100 points/ addresses. A feature allowing path and polygon annotations, which can be exported to KML, was formerly only available to Plus users, but was made free in version 4.0.2416.
- Higher data download speeds

Google Earth Pro

Google Earth Pro is a business-oriented upgrade to Google Earth that has more features than the Plus version. It is the most feature-rich version of Google Earth available to the public, with various additional features such as a movie maker and data importer. In addition to business-friendly features, it has also been found useful for travelers with map-making tools. Up until late January 2015, it was available for $399/year, however Google decided to make it free to the public. It is now for free and Google does not mention anything about new policy changes. The Pro version includes add-on software such as:

- Movie making.
- GIS data importer.
- Advanced printing modules.
- Radius and area measurements.

Google Earth Pro is available for Windows (NT-based versions), Mac OS X 10.4 or later.

Google Earth Enterprise

Google Earth Enterprise is a version of Google Earth designed for use by organizations whose businesses could take advantage of the program's capabilities, for example by having a globe that holds company data available for anyone in that company. As of March 20, 2015, Google has retired the Google Earth Enterprise product, with support ending March 22, 2017.

Automotive Version

An automotive version of Google Earth is available in the 2010 Audi A8.

Google Earth Plug-in

The Google Earth API is a free beta service, available for any web site that is free to consumers. The Plug-in and its JavaScript API let users place a version of Google Earth into web pages. The API enables sophisticated 3D map applications to be built. At its unveiling at Google's 2008 I/O developer conference, the company showcased potential applications such as a game where the player controlled a milktruck atop a Google Earth surface.

The Google Earth API has been deprecated as of 15 December 2014 and will remain supported until the 15th of December 2015. Google Chrome aims to end support for the Netscape Plugin API (which the Google Earth API relies on) by the end of 2016.

Controversy and criticism

The software has been criticized by a number of special interest groups, including national officials, as being an invasion of privacy and even posing a threat to national security. The typical argument is that the software provides information about military or other critical installations that could be used by terrorists.

- Former President of India APJ Abdul Kalam expressed concern over the availability of high-resolution pictures of sensitive locations in India. Google subsequently agreed to censor such sites.
- The Indian Space Research Organization said Google Earth poses a security threat to India, and seeks dialogue with Google officials.
- The South Korean government expressed concern that the software offers images of the presidential palace and various military installations that could possibly be used by hostile neighbor North Korea.
- In 2006, one user spotted a large topographical replica in a remote region of China. The model is a small-scale (1/500) version of the Karakoram Mountain Range, which is under the control of China but claimed by India. When later confirmed as a replica of this region, spectators began entertaining military implications.
- In 2006, Google Earth began offering detailed images of classified areas in Israel. The images showed Israel Defense Forces bases, including secret Israeli Air Force facilities, Israel's Arrow missile defense system, military headquarters and Defense Ministry compound in Tel Aviv, a top-secret power station near Ashkelon, and the Negev Nuclear Research Center. Also shown was the alleged headquarters of the Mossad, Israel's foreign intelligence service, whose location is highly classified.
- Operators of the Lucas Heights nuclear reactor in Sydney, New South Wales, Australia asked Google to censor high-resolution pictures of the facility. However, they later withdrew the request.

- In July 2007, it was reported that a new Chinese Navy Jin-class nuclear ballistic missile submarine was photographed at the Xiaopingdao Submarine Base south of Dalian.
- Hamas and the al-Aqsa Martyrs' Brigades have reportedly used Google Earth to plan Qassam rocket attacks on Israel from Gaza (See: List of Qassam rocket attacks.)
- The lone surviving gunman involved in the 2008 Mumbai attacks admitted to using Google Earth to familiarise himself with the locations of buildings used in the attacks.
- Michael Finton, aka Talib Islam, used Google Earth in planning his attempted September 24, 2009, bombing of the Paul Findley Federal Building and the adjacent offices of Congressman Aaron Schock in Springfield, Illinois.
- In 2009, Google superimposed old woodblock prints of maps from 18th and 19th century Japan over Japan today. These maps marked areas inhabited by the burakumin caste, who were considered "non-humans" for their "dirty" occupations, including leather tanning and butchery. Descendants of members of the burakumin caste still face discrimination today and many Japanese people feared that some would use these areas, labeled *etamura* (*translation: "village of an abundance of defilement""*), to target current inhabitants of them. These maps are still visible on Google Earth, but with the label removed where necessary.

Countries where Google Earth is Blocked:				
Country	**By Whom**	**Reason**	**Since When**	**Source**
Iran	Google	US government export restrictions	2007	
Morocco	Maroc Telecom, the most popular service provider	Unknown	2006	
Sudan	Google	US government export restrictions	2007	

Google Earth has been blocked by Google in Iran and Sudan since 2007 due to US government export restrictions. The program has also been blocked in Morocco since 2006 by Maroc Telecom, a major service provider in the country.

Blurred out image of the Royal Stables in The Hague, Netherlands. This has since been partially lifted.

Some citizens may express concerns over aerial information depicting their properties and residences being disseminated freely. As relatively few jurisdictions actually guarantee the individual's right to privacy, as opposed to the state's right to secrecy, this is an evolving point. Perhaps aware of these critiques, for a time, Google had Area 51 (which is highly visible and easy to find) in Nevada as a default placemark when Google Earth is first installed.

As a result of pressure from the United States government, the residence of the Vice President at Number One Observatory Circle was obscured through pixelization in Google Earth and Google Maps in 2006, but this restriction has since been lifted. The usefulness of this downgrade is questionable, as high-resolution photos and aerial surveys of the property are readily available on the Internet elsewhere. Capitol Hill also used to be pixelized in this way. The Royal Stables in The Hague, Netherlands also used to be pixelized, and are still pixelized at high zoom levels.

Critics have expressed concern over the willingness of Google to cripple their dataset to cater to special interests, believing that intentionally obscuring any land goes against its stated goal of letting the user "point and zoom to any place on the planet that you want to explore".

In the United Kingdom, critics have also argued that Google Earth has led to the vandalism of private property, highlighting the graffiti of a penis being drawn on the roof of a house near Hungerford, on the roof of Yarm School at Stockton on Tees and on the playing fields of a school in Southampton as examples of this.

In Hazleton, Pennsylvania, media attention and critics focused on Google Earth once more because of the defacing of the Hazleton Area High School football field. Grass was removed to create the image of a penis approximately 35 yards long and 20 yards wide.

Late 2000s versions of Google Earth require a software component running in the background that will automatically download and install updates. Several users expressed concerns that there is not an easy way to disable this updater, as it runs without the permission of the user.

In the academic realm increasing attention has been devoted to both Google Earth and its place in the development of digital globes more generally. In particular, the International Journal of Digital Earth now features many articles evaluating and comparing the development Google Earth and its differences when compared to other professional, scientific and governmental platforms.

Elsewhere, in the Humanities and Social Sciences, Google Earth's role in the expansion of "earth observing media" has been examined. Leon Gurevitch in particular has examined the role of Google Earth in shaping a shared cultural consciousness regarding climate change and humanity's capacity to treat the earth as an engineerable object. Gurevitch has described this interface between earth representation in Google Earth and a shared cultural imaginary of geo-engineering as "Google Warming".

Copyright

Every image created from Google Earth using satellite data provided by Google Earth is a copyrighted map. Any derivative from Google Earth is made from copyrighted data which, under United States Copyright Law, may not be used except under the licenses Google provides. Google allows non-commercial personal use of the images (e.g. on a personal website or blog) as long as copyrights and attributions are preserved. By contrast, images created with NASA's globe software World Wind use The Blue Marble, Landsat or USGS layer, each of which is a terrain layer in the public domain. Works created by an agency of the United States government are public domain at the moment of creation. This means that those images can be freely modified, redistributed and used for commercial purposes.

Layers

Google Earth also features many layers as a source for information on businesses and points of interest, as well as showcasing the contents of many communities, such as Wikipedia, Panoramio and YouTube.

Borders and Labels

Contains borders for countries/provinces and shows placemarks for cities and towns.

- Borders: *Marks international borders with a thick yellow line (borders with territorial disputes with thick red lines), 1st level administrative borders (generally provinces and states) with a lavender line, and 2nd level administrative borders (counties) with a cyan line. Coastlines appear as a thin yellow line. Displays names of countries, 1st level administrative areas, and islands.*
- Labels: *Displays labels for large bodies of water, such as oceans, seas, and bays, and populated places.*

3D Imagery

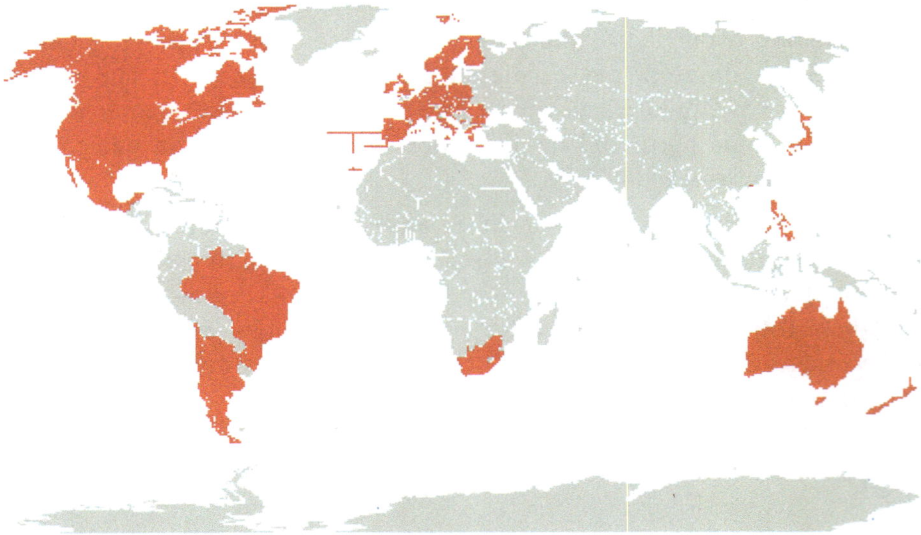

3D coverage in Google Earth

Google Earth 3D shows many 3D computer graphics building models in many cities, in these styles:

- 3D trees: Shows many trees in Athens, Greece; Surui Forest, Brazil; Kahigaini, Kenya; Mangrove Forests, Mexico; Jedediah Smith Redwoods State Park, California
- Photorealistic: Shows many buildings in a realistic style, with more complex polygons and surface images.
- Autogen: Renders entire metropolitan areas in 3D via processing of 45 degree aerial imagery.
- Gray: Low-detail models of city buildings designed for computers that may not have the capability of showing the photorealistic models.

In 2009, in a unique collaboration between Google and the Museo del Prado in Madrid, the museum selected 14 of its most important paintings to be photographed and displayed at the ultrahigh resolution of 14,000 megapixels inside the 3D version of the Prado in Google Earth and Google Maps.

In June 2012, Google announced that it will be replacing user made 3D buildings with an auto-generated 3D mesh. This will be phased in, starting with select larger cities, with the notable exception of cities such as London and Toronto which require more time to process detailed imagery of their vast number of buildings. The reason given is to have greater uniformity in 3D buildings, and to compete with other platforms already using the technology such as Nokia Here and Apple Maps. The first 3D buildings in Google Earth were created using 3D modeling applications such as SketchUp and, beginning in 2009, Building Maker, and were uploaded to Google Earth via the 3D Warehouse.

In 2012, Google began incorporating automatically-generated 3D imagery, which displays entire areas in 3D rather than individual buildings, into the mobile and desktop versions of Google Earth, releasing coverage of 21 cities in four countries that year. By March 2015, 3D imagery covering more than 300,000 km^2 was available and by early 2016 had been expanded to hundreds of cities in over 40 countries, including every U.S. state and encompassing every continent except Antarctica.

During 2015, Hong Kong and places in the Philippines were added to the coverage.

As of February 2016, 3D imagery covering more than 495,000 km^2 was available in cities in over 40 countries, covering all continents except Antarctica.

Google Street View

Shows placemarks with 360 degree panoramic views of streets of many cities in Australia, France, the United Kingdom, Republic of Ireland, Italy, Japan, New Zealand, Spain, the United States, and recently Portugal, Brazil, the Netherlands, Taiwan, Switzerland, Canada, Mexico, Sweden, Norway, South Africa and Finland.

Weather

- Clouds – *Displays cloud cover based on data from both geostationary and low Earth-orbiting satellites. The clouds appear at their calculated elevation, determined by measuring the cloud top temperature relative to surface temperature.*
- Radar – *Displays weather radar data provided by weather.com and Weather Services International, updating every 5–6 minutes.*
- Conditions and Forecast – *Displays local temperatures and weather conditions. Clicking on an indicator displays a 2 Day Forecast (Example: Monday Morning, Monday Night, Tuesday Morning, Tuesday Night) forecast provided by weather.com.*
- Information – *Clicking Information allows users to further read up on where Google Earth gets weather information.*

Sky layers

Layers for Google Sky.

- EarthSky Podcasts

GRASS GIS

Geographic Resources Analysis Support System (commonly termed *GRASS GIS*) is a geographic information system (GIS) software suite used for geospatial data management and analysis, image processing, producing graphics and maps, spatial and temporal modeling, and visualizing. It can handle raster, topological vector, image processing, and graphic data.

GRASS GIS contains over 350 modules to render maps and images on monitor and paper; manipulate raster and vector data including vector networks; process multispectral image data; and create, manage, and store spatial data.

It is licensed and released as free and open-source software under the GNU General Public License (GPL). It runs on multiple operating systems, including OS X, Windows and Linux. Users can interface with the software features through a graphical user interface (GUI) or by *plugging into* GRASS via other software such as QGIS. They can also interface with the modules directly through a bespoke shell that the application launches or by calling individual modules directly from a standard shell. The latest stable release version (LTS) is GRASS GIS 7, which is available since 2015

The GRASS Development Team is a multinational group consisting of developers at many locations. GRASS is one of the eight initial Software Projects of the Open Source Geospatial Foundation.

Architecture

GRASS supports raster and vector data in two and three dimensions. The vector data model is topological, meaning that areas are defined by boundaries and centroids; boundaries cannot overlap within one layer. In contrast, OpenGIS Simple Features, defines vectors more freely, much as a non-georeferenced vector illustration program does.

GRASS is designed as an environment in which tools that perform specific GIS computations are executed. Unlike GUI-based application software, the GRASS user is presented with a Unix shell containing a modified environment that supports execution of GRASS commands, termed modules. The environment has a state that includes parameters such as the geographic region covered and the map projection in use. All GRASS modules read this state and additionally are given specific parameters (such as input and output maps, or values to use in a computation) when executed. Most GRASS modules and abilities can be operated via a graphical user interface (provided by a GRASS module), as an alternative to manipulating geographic data in a shell.

The GRASS distribution includes over 350 core modules. Over 100 add-on modules created by users are offered on its website. The libraries and core modules are written in C. Other modules are written in C, C++, Python, Unix shell, Tcl, or other scripting languages. The modules are designed under the Unix philosophy and hence can be combined using Python or shell scripting to build more complex or specialized modules, by users, without knowledge of C programming.

There is cooperation between the GRASS and Quantum GIS (QGIS) projects. Recent versions of QGIS can be executed within the GRASS environment, allowing QGIS to be used as a user-friendly graphical interface to GRASS that more closely resembles other graphical GIS software than does the shell-based GRASS interface.

Another project exists to re-implement GRASS in Java as *JGRASS*.

REMOTE VIEW

RemoteView is the family name of a group of software programs designed to aid in analyzing satellite or aerial images of the Earth's surface for the purpose of collecting and disseminating geospatial intelligence. The National Geospatial-Intelligence Agency (NGA) is a longtime user of RemoteView software. The software is published by Textron

Systems Geospatial Systems.

RemoteView is a widely available electronic light table application, initially developed and released commercially by Sensor Systems in 1996. An electronic light table application makes it possible for imagery analysts to review satellite images on a computer instead of examining film or printed photographs. RemoteView was originally written only for the Unix operating system, but as the US Department of Defense transitioned to the Windows operating system, Sensor Systems released a Windows-based version that soon became a primary tool used by the intelligence community. Overwatch acquired Sensor Systems and the RemoteView software in 2005. Textron Systems acquired Overwatch in 2006.

RemoteView's main function is an imagery and geospatial analysis tool. It can display a wide array of imagery formats, elevation data sets, and vector data sets. Capabilities include image enhancements, photogrammetry, orthorectification, multispectral classification, pan sharpening, change detection, assisted search, location positioning, and 3D terrain visualization. These features allow an intelligence analyst to review large-scale imagery and generate annotated reports on any findings.

RemoteView for Windows comes in two versions, RemoteView Desktop and RemoteView Pro. RemoteView Desktop is an entry-level imagery exploitation and analysis solution, for focusing on a single image at a time or reviewing the results of another analyst's work. RemoteView Pro is the flagship product offering a greater number of tools, extensions, and capabilities.

TerrSet

TerrSet (formerly IDRISI) is an integrated geographic information system (GIS) and remote sensing software developed by Clark Labs at Clark University for the analysis and display of digital geospatial information. TerrSet is a PC grid-based system that offers tools for researchers and scientists engaged in analyzing earth system dynamics for effective and responsible decision making for environmental management, sustainable resource development and equitable resource allocation.

Key features of TerrSet include:

- a complete GIS analysis package for basic and advanced spatial analysis, including tools for surface and statistical analysis, decision support, land change and prediction, and image time series analysis;
- a complete Image Processing system with extensive hard and soft classifers, including machine learning classifiers such as neural networks and classification tree analysis, as well as image segmentation for classification;
- Land Change Modeler, a land planning and decision support toolset that addresses the complexities of land change analysis and land change prediction.
- Habitat and Biodiversity Modeler, a modeling environment for habitat assessment and biodiversity modeling.
- Ecosystem Services Modeler, a spatial decision support system for assessing the value of natural capital.

- Earth Trends Modeler, an integrated suite of tools for the analysis of image time series to assess climate trends and impacts.
- Climate Change Adaptation Modeler, a facility for modeling future climate and its impacts.
- GeOSIRIS-REDD, a national-level REDD planning tool to assess deforestation, carbon emissions, agricultural revenue and carbon payments.

COGNITION NETWORK TECHNOLOGY

Cognition Network Technology (CNT, also known as Definiens Cognition Network Technology) is an object-based image analysis method developed by Nobel laureate Gerd Binnig together with a team of researchers at Definiens AG in Munich, Germany. It serves for extracting information from images using a hierarchy of image objects (groups of pixels), as opposed to traditional pixel processing methods.

To emulate the human mind's cognitive powers, Definiens used patented image segmentation and classification processes, and developed a method to render knowledge in a semantic network. CNT examines pixels not in isolation, but in context. It builds up a picture iteratively, recognizing groups of pixels as objects. It uses the color, shape, texture and size of objects as well as their context and relationships to draw conclusions and inferences, similar to a human analyst.

Company

In 1994 Professor Gerd Binnig founded *Definiens*. CNT was first available with the launch of the *eCognition* software in May 2000. In June 2010, Trimble Navigation Ltd (NASDAQ: TRMB) acquired Definiens business asset in earth sciences markets, including eCognition software, and also licensed Definiens' patented CNT. In 2014, Definiens was acquired by MedImmune, the global biologics research and development arm of AstraZeneca, for an initial consideration of $150 million.

Software

Definiens Tissue Studio

Definiens Tissue Studio is a digital pathology image analysis software application based on CNT. The intended use of Definiens Tissue Studio is for biomarker translational research in formalin-fixed, paraffin-embedded tissue samples which have been treated with immunohistochemical staining assays, or hematoxylin and eosin (H&E).

The central concept behind Definiens Tissue Studio is a user interface that facilitates machine learning from example digital histopathology images in order to derive an image analysis solution suitable for the measurement of biomarkers and/or histological features within pre-defined regions of interest on a cell-by-cell basis, and within sub-cellular compartments. The derived image analysis solution is then automatically applied to subsequent digital images in order to objectively measure defined sets of multiparametric image features. These data sets are used for further understanding the underlying biological processes that drive cancer and other diseases. Image processing and data analysis are performed either on a local desktop computer workstation, or on a server grid.

eCognition

The *eCognition* suite offers three components which can be used stand-alone or in combination to solve image analysis tasks. eCognition Developer is a development environment for object-based image analysis. It is used in earth sciences to develop rule sets (or applications) for the analysis of remote sensing data. eCognition Architect enables non-technical users to configure, calibrate and execute image analysis workflows created in eCognition Developer. eCognition Server software provides a processing environment for batch execution of image analysis jobs.

eCognition software is utilized in numerous remote sensing and geospatial application scenarios and environments, using a variety of data types:

- Generic: Rapid Mapping, Change Detection, Object Recognition
- By environment: Diverse Landcover Mapping, Urban Analysis (i.e. impervious surface area analysis for taxation, property assessment for insurance, inventory of green infrastructure), Forestry (i.e. biomass measurement, species identification, firescar measurement), Agriculture (i.e. regional planning, precision farming, crisis response), Marine and Riparian (i.e. ecosystem evaluation, disaster management, harbor monitoring).
- Other: Defense, security, atmosphere and climate

The online eCognition community was launched in July 2009 and had 2813 members as of July 9, 2010. Membership is distributed globally and user conferences are held regularly, the last having taken place in November 2009 in Munich, Germany. The bi-annual GEOBIA (Geographic Object-Based Image Analysis) conference is heavily attended by eCognition users, with the majority of presentations based on eCognition software.

REFERENCES

1. Campbell, J.B. (2002). Introduction to remote sensing (3rd ed.). The Guilford Press. ISBN 1-57230-640-8.
2. Jensen, J.R. (2007). Remote sensing of the environment: an Earth resource perspective (2nd ed.). Prentice Hall.
3. Jensen, J.R. (2005). Digital Image Processing: a Remote Sensing Perspective (3rd ed.). Prentice Hall.
4. Lillesand, T.M.; R.W. Kiefer; J.W. Chipman (2003). Remote sensing and image interpretation (5th ed.). Wiley. ISBN 0-471-15227-7.
5. Richards, J.A.; X. Jia (2006). Remote sensing digital image analysis: an introduction (4th ed.). Springer. ISBN 3-540-25128-6.

6. Rencz, A.N. (1999). Remote Sensing for the Earth Sciences: Manual of Remote Sensing (3rd ed.). Wiley. ISBN 0-471-29405-5.

Index

L

M

O

P

R

S

T

U

W